Phenomenology in Japan

Edited by

ANTHONY J. STEINBOCK

Department of Philosophy, Southern Illinois University at Carbondale, USA

Reprinted from *Continental Philosophy Review*, Volume 31 (3), 1998

Kluwer Academic Publishers

Dordrecht / Boston / London

A.I.P. Catalogue record for this book is available from the Library of Congress.

ISBN 978-90-481-5118-9

Published by Kluwer Academic Publishers,
P.O. Box, 3300 AA Dordrecht, The Netherlands.

Sold and distributed in North, Central and South America
by Kluwer Academic Publishers,
101 Philip Drive, Norwell, MA 02061, U.S.A.

In all other countries, sold and distributed
by Kluwer Academic Publishers,
P.O. Box 322, 3300 AH Dordrecht, The Netherlands.

Printed on acid-free paper

TABLE OF CONTENTS

For my Mother, Marie Steinbock

We must now declare that because "dim eyesight" is equal, the "flowers of emptiness" are equal; that because "dim eyesight" is birthless, the "flowers of emptiness" are birthless; and that because all dharmas are real nature, the "flowers of dimness" are real nature. These do not concern themselves with past, present, or future, nor do they have anything to do with the beginning, middle, or end. Because they are not obstructed by arising or perishing, they freely cause arising and perishing to arise and perish. They arise in "emptiness" and perish in "emptiness"; they arise in "dimness" and perish in "dimness"; they arise in "flowers" and perish in "flowers." They do this at all times and in all places.

From Dōgen's Shōbōgenzō/Kūge (trans. Hee-Jin Kim)

I mean now: We normally live in our homeworld, or better, in an environing-world which is a world actually familiar to us (albeit a world which is not familiar in all individual realities), a world actually to be realized for us in intuition. In the mediate horizon are heterogenous humanities and cultures; they belong to them as alien and as heterogenous; but alienness means accessibility in genuine inaccessibility, in the mode of incomprehensibility.

Edmund Husserl (1933)

Continental Philosophy Review **31**: 225–238, 1998.
© 1998 *Kluwer Academic Publishers.*

Introduction: Phenomenology in Japan

ANTHONY J. STEINBOCK
*Department of Philosophy, Southern Illinois University at Carbondale, Carbondale,
IL 62901-4505, USA*

1. Phenomenology in Japan and the prospect of cross cultural communication

The very title of this collection of essays, "Phenomenology in Japan" signals an effort that might well be called an attempt at cross cultural communication or cultural interchange. It might evoke curiosity or interest: *"Phenomenology in Japan!"* (as if to say, "how is phenomenology being done in Japan?" or "let's see what the Japanese have to say about phenomenology!"); or it can be met with surprise: "Phenomenology, *in Japan?*" (as if to inquire, "are philosophers and social scientists really occupied with phenomenology in Japan?" or "can one really speak of a 'phenomenological movement' in Japan?").

In either case there is a certain assumption in play, namely, that phenomenology has traditionally belonged to the West, especially Europe, and later America (which, for Husserl, still belonged to the spirit of Europe), and having had its roots there is expressive of a certain style of thinking that is peculiarly "Western" and alien to the "East." Indeed, a collection that would be entitled and devoted to "Phenomenology in Europe" would not pique the same sort of interest or directly raise the same sort of issues, and would almost seem trivial, whereas "Phenomenology in Japan" suggests something different, something worthy of attention, at least to those in the West. Moreover, and for reasons that concern the very nature of the matter at hand, a publication in Japan that would feature something like "Buddhism in America" could not be understood as a symmetrical inversion of "Phenomenology in Japan." These are important experiential clues that we should not take lightly. They are rooted in a clash and irreversibility between the eventual insight into emptiness or *sunyata* as it emerged from a particular tradition within Buddhism (emptiness which is itself no-thing, not being and not not-being, either), and the insight into the generative structure of experience of the West that emphasizes the generation and orientation of meaning(s) as historically, personally, and ontologically significant.

[1]

Historically speaking, phenomenology was introduced in Japan in the mid-1900s by Kitaro Nishida (1870-1945). While he had a strong background in the Western history of philosophy and often used problems from the West as his point of departure, his thinking was executed in a way and with a tension that allowed him to remain "Eastern." This is rooted not only in his Buddhist background and practice, but also in his original and profound insights into reality as the self-identity of absolute contradictories.[1]

Aside from Nishida's attentiveness to Husserl's *Logische Untersuchungen* and his *Ideen* in the 1910s, the introduction of phenomenology to Japan was continued in the 1920s and 1930s by Japanese thinkers who studied with Husserl, Heidegger, Fink, and Becker, philosophers like Hajime Tanabe, Satomi Takahashi, Tetsuro Watsuji, Goichi Miyake as well as phenomenological sociologists like Tomoo Otaka, Kazuta Kurauchi, and Jisho Usui.[2]

Presently the "Phenomenological Association of Japan" (NIHON GEN-SHOUGAKUKAI) which was founded in 1979, counts approximately 400 members, and is one of the largest philosophical associations next to the more general "Philosophical Association of Japan (NIHON TETSUGAKUKAI)." The Phenomenological Association is joined by "The Association of Phenomenology and Hermeneutics (GENSHOUGAKU KAISHAKUGAKU KENKYUUKAI)" and "The Association of Phenomenology and Social Sciences (NIHON GENSHOUGAKU SHAKAIKAGAKUKAI)," each of which has about 100 members. This growing interest in phenomenology is reflected in recent special and regular conferences that feature phenomenological styles of inquiry, and publications like *Japanische Beiträge zur Phänomenologie*,[3] *Phänomenologie der Praxis im Dialog zwischen Japan und dem Western*,[4] *Japanese and Western Phenomenology*,[5] and "Phenomenology and the Human Sciences in Japan".[6]

I mentioned certain experiential clues above that bear on an interfacing of traditions as radically distinct as East and West. Naturally, these clues do not merely concern phenomenology or philosophy taken in the narrow sense. In his film, *Kagemusha*, staged in 16th century Japan, Kurosawa highlights an interfacing that has profound cultural and religious significance. For Kurosawa, Japan became alien to itself in a drastic manner through the appropriation – uncritical appropriation – of Western values. There are many clues in this direction such as the savoring of imported European wine in an encampment, the adoption of new military technology from the West, but perhaps most poignantly the relinquishing of a particular style of battle that had been the hallmark of a particular clan. This style is captured by the dictum "to remain still as the mountain." After the death of the warlord Shingen Takeda, and the banishment of his double who replaced him for three years after the former's death, the leadership of the Takeda clan falls to Shingen's son, Katsuyori. Katsuyori betrays this style of fighting and style of being, and rather than remaining still like the mountain, actively moves out to attack smaller clans, which ultimately brings his own clan's defeat.

In this film, Kurosawa is attempting to take a critical, i.e., reflective stance in relation to what we often take for granted today, namely, intercultural encounters and the implicit or even explicit appropriation of systems of value. The contributors to this special collection also take up a critical, i.e., reflective or *phenomenological* perspective to address – even if in an indirect way – a predominant feature that gets imported from the West and qualifies existence in the world itself: "dualism." Interesting to me as editor of this edition is that none of the contributors were asked to address the issue of dualism; rather, they spontaneously touched on this issue in one way or another, even if it was not their main focus. If Toru Tani's assertions in his essay are correct, the Japanese have become so permeated with a Western style of subject-object thinking that it has become their own starting point even when they are criticizing it, and this he writes makes even the thought of a Japanese philosopher like Nishida alien to themselves.[7]

Also interesting is the fact that the problem of dualism is being addressed through *phenomenology*. This is perhaps less curious since the authors' main work is in phenomenology. But the question still remains why phenomenology, among all the other Western styles of thinking, could be seen as capable of handling this kind of problem, and why its results, tentative as they may be, can be fruitful in this regard. One reason for this, I think, can be stated as follows: Even though – or precisely because! – phenomenology is a product of the West, there are resources within phenomenology that help point beyond problems that a Western style of thinking implicitly brings with it. And this would apply to the *de facto* problems of dualism that the East itself faces in its own way. Specifically, instead of imposing theoretical systems upon experience, constructing experience according to predetermined categories, phenomenology liberates the phenomena and thereby clarifies that experience which is inherently multivalent, over-determined, or as Merleau-Ponty might say, "ambiguous." Even though phenomenology as a style of approaching the phenomena did at first run into a few of its own snags and methodological stumbling blocks, it did not remain dogmatic, but retained a kind of sincerity in its approach and an honesty to its descriptions. This may indeed account for how, through its self-transformations, it is able to elucidate an incredibly wide range of experiential phenomena. And this may also account for the various lines of convergence that many Easterns and Westerns find between certain strains of phenomenology and Buddhist thinking.

I will discuss the main points of the author's contributions below, the issues that they target as important, and their way of grappling with them through phenomenology. Here I would like to return to the issues mentioned above and to pose a question that bears on their very structure: How can the foreign or the alien become integrated into a way of thinking such that it can survive as both a presupposition of thought and as alien to thought? Or viewed more specifically even in terms of the title of this special issue, granted that phenomenology in Japan is active, thriving, and even irreplaceable, how is it possible to speak of phenomenology in Japan without presupposing the mere fact of phenomenology in Japan?

2. Becoming alien of the home and the generative structure of experience

Max Scheler has accounted for the possibility of incommensurate world-views by what he calls the "functionalization of essence."[8] By functionalization of essence, Scheler means a process by which the very being of things guides our take on things, ultimately, the way in which essential structures guide our apprehension of reality. This process, viewed historically and intersubjectively, stylizes and typifies the very way of "seeing" and the very reality seen. Eventually, it would be possible to speak of cultures and groups of peoples that, on the one hand, share basic presuppositions of reality, and on the other, have *different* paths of access from the realm of facts to the essential structures of reality, even those that are radically incommensurate: for example, those that have cultivated the insight into generativity, and those that have cultivated the insight into emptiness. Not only would it be essentially impossible to mask these differences or to substitute one path of access for another, as if the different worlds and world-views would be reversible, but according to Scheler, if a particular "path of access" were lost, e.g., by a people being annihilated, culturally assimilated, etc., these insights might not ever be gained again, not just factually, but in principle, due to the loss of the generative density of that unique and irreducible way of seeing and living. Scheler's point is not that we cannot and do not share cultural invariants, but that given this dynamic, generative structure, we might be *unable* to see what other peoples and other ages have seen. This points not only to the necessity for collaborative inquiry where epistemology is concerned, but for the love of others in solidarity where the religious and moral life is concerned. Not the incommensurability, but the loss of a way of seeing would constitute a diminution in the spiritual growth of humanity.

In his later years, Husserl – the so-called founder of phenomenology – had a way of getting at this sense of intercultural phenomena through the description of socially, geohistorically, and normatively significant lifeworlds, or what he called within a *"generative"* framework, the interrelations of "homeworlds" and "alienworlds." The home is not one place among others, but a normatively special geo-historical place that is constituted with a certain asymmetrical privilege, and it can range from the smallest generative unit, "mother or parents and child," to a virtual cultural world. The home gets this asymmetrical privilege through modes of appropriation and disappropriation of sense that extend historically over the generations. Along the lines of Scheler, these modes of appropriation and disappropriation express particular styles of access to reality, the ways of being guided by essential structures, and the ways in which connections are made, etc. For both Scheler and Husserl, these are selective/exclusive, a process that Husserl calls "optimalization." The modes of accessibility and inaccessibility are constituted and transmitted through such things as ritual, narrative, language, shared habits and customs, styles of movement and thinking, and so forth, that bring the essential structures to bear in this way rather than that. Accordingly, what gets constituted as "home" is not only a "ground-horizon" as a basis for living, but the very lifeworld to which we return. The home becomes normatively significant to us as experience gets shaped concordantly and optimally, and over time typically, and with familiarity. In this way, Husserl understands the home to be constituted for the "homecompanions," generally speaking, in the mode of "accessibility."

Though constituted with an experiential and normative weight through the appropriative/disappropriative process as "our," the home however is not a one-sided, independent original sphere independent of the alien. Through the generative constitution of the sense of "home," an alienworld or alienworlds are liminally co-constituted *as* alien, in the extreme case, *as* neither concordant, nor optimal, nor typical, nor familiar. Accordingly, Husserl understands alienness to be constituted precisely as accessibility in the mode of genuine inaccessibility and incomprehensibility. Through modes of appropriation and disappropriation of the home, which is always in the process of being constituted as home, the alien is co-constituted as alien. Moreover, not only is the constitution of the home liminally co-constitutive of the alien, but the alien is co-constitutive of the home through modes of transgressive experience.

By transgressive experience I understand a relation with the alien that crosses over the "limits" of the home, but from within the home, and such that the limits of the home are only exposed in the encounter with the alien, and are never encountered like an object. According to Husserl, the home does not exist as an independent sphere of ownness, but is only constituted as home through the alienworld(s). Transgression is the process of crossing over the limits of the home while remaining rooted in the home, and thus bringing an explicit experience of limits into being. The generative relation of home/alien that has been invoked here is itself not a thing to be encountered; rather, it *emerges as such through the encounter* of the home with the alien, through liminal experience. Through the alien, we gain the home as home. The structure, home/alien, is a co-original, co-foundational structure: Because we are constituted as "home," we belong to the alien in the process of co-constitution, but precisely as not belonging to the alien as being home; and because we can encounter the alien through processes that transgress the limits of the home from within the home, we encounter the home as if "for the first time," through the encounter of the alien. In short, the structure, home/alien exhibits concretely varying degrees of homeness and alienness, since it is co-constituted through an optimalizing process that Husserl calls "generativity."[9]

Two important interrelated considerations arise concerning the sense of a collection like this one, "Phenomenology in Japan," and bear on many of the themes explicitly and implicitly treated by the contributors to this special issue. First, to what extent does the home remain home and to what extent is it susceptible to the alien? Second, is the generative relation of home/alien an essential structure that holds across cultures, or is this "essential structure" itself only peculiar to the West?

In the first place, because the home is a *liminal* formation (that is, is co-constituted as home precisely through its de-limitation from the alien), there is a *becoming alien of the home*. This becoming alien of the home belongs to the very structure of home/alien. In fact, the alien could not be encountered *as alien* in the mode of genuine inaccessibility and incomprehensibility if the home were hermetically sealed off from the alien. The alien as alien is encountered as alien from the home. Moreover, there could be no encounter with the alien if it putatively occurred within the alien. If the latter were the case, this so-called "alien" could never have the sense as "alien," "strange," or "foreign," nor could the home be constituted as home.

The very same thing that allows the alien to be encountered as alien (as unfamiliar, as a-typical, and hence as not home) is the very same thing that opens the home to the influence of the alien. Since the home only opens up as home from the liminal encounter with the alien, alienness must seep into the very experience peculiar to the home; not only is there a becoming other through the encounter of others within the home, there is also a transformation of the home through the alien. Thus, Husserl writes in a text from the early 1930s, that we have a transformation of my and our world-experience and a transformation of the world itself through the encounter with the alien (Hua XV 214), and further that it belongs to the very general type of the being of the homeworld that the concrete particular homeworld is transformed through encounters with the alien (Hua XV 222), encounters that I have referred to above as "liminal."[10]

This presentation of the structure of home/alien allows us to understand basic conditions not only of how phenomenology can itself become integrated into the homeworld of thinking in Japan, but also of how a *style* of thinking (not just objects of thought) could become so integrated into a foreign way of thinking that it actually becomes, at least to a large extent, the "home" style of thinking for that world. But there is another question raised above concerning whether the structure home/alien is a cross cultural structure, or if in its incommensurability it is "only" peculiar to the West. I will suggest that from a generative phenomenological perspective the structure home/alien is both a structure of the "whole," and it is only peculiar to the West. Let me explain.

First, generativity was indeed "discovered" in the West and is the very process by which there are normatively significant structures that have a unique and irreducible orientation, and that through their difference, make a difference, permitting not only the experiences of anticipation, disappointment, crises, but also of overcoming them. As noted, generativity becomes articulated normatively, socially, and geo-historically in the very structure home/alien. When we speak of the generative framework as home/alien, we are describing the movement of generativity, and hence the "whole" generative framework. The whole generative framework, however, is not described from an objective, third person perspective, but from within the home, in this case, within generativity. Accordingly, the structure of generativity as I have expressed it here does not merely account for differences that would be alien to a particular home, but for the possibility of something radically alien even to generativity itself!

When speaking of the "whole" structure from within a generative perspective, we are placed in the peculiar situation of describing the whole from within the home as in relation to the alien. In this respect, generativity is a structure of the whole. But directly related to this, and for the same reasons, one must also say that the generative structure of home/alien arises from the insight into reality as generativity, and is thus peculiar to the West. Thus, from the Western point of view, generativity takes the form of home/alien and "defines" this perspective as "home." This is the whole structure interpersonally and historically clarifying itself in terms of home/alien. For the East, however, the so-called whole structure is clarified, in Nishida's terms, as the self-identity of absolute contradictories. On the one hand this means the East in relation to the West; but its "point" would be emptiness, whereas for the West it would be generativity.[11]

The challenge here, for us in the West, is having to speak of the whole generative framework expressed in terms of homeworld and alienworld from within the perspective of the home *but without resolving the tension of home/alien and thus closing off the unique modes of expression peculiar to the alien which may call the home (e.g. generativity) into question.* This is further complicated by the fact that for us the *only access to the whole is precisely in the encounter with the alien within generativity*: the generative framework is given only in this incongruous, absolutely irreducible relation and not outside of it. Because we bring the generative density of the home with us, we speak through the home toward the alien. This exemplifies the structure of "transgressive experience" that I alluded to above.

Cross cultural communication does not have to entail melding my background with that of others in order to communicate, for the communication takes place precisely through the generative differences of home and alien. Since the constitution of the generative framework is both a co-constitution of the alien through appropriative experience of the home, and as the co-constitution of the home through the transgressive experience of the alien, the goal of cross cultural communication can*not* in the first instance be that the alien understand me, but that by developing the position of the home, the alienworld is thrown back on itself and understands itself more deeply. This encounter with the alien from the perspective of the home, then, allows the home to be critical with respect to itself.

I have described generativity as the generation of historically significant meaning that is expressed as the irreducible and irreversible relation of home-world and alienworld. If generativity is to be sensitive to its own generative situatedness, it cannot take itself for granted, and it cannot address generativity as if, for example, the problem of "emptiness" were of no consequence to the East and as if emptiness could simply be integrated by generativity.

This is not the same as asserting, as a Westerner, that generativity is simply a "narrative" of the West. To put it forth as one narrative among others would be *to relativize* the home (and the alien) and to presuppose that I could somehow abstract the home from the relation, comparing it to the East by some overarching supposed neutral term. Instead, it is precisely in the face of emptiness that generative phenomenology can describe generativity, for to communicate generativity cross culturally demands doing so within generativity in the face of emptiness. Cross cultural communication as a crossing over from within entails describing the generative framework *fully* from the home as it is open to being called into question by the alien in and through the liminal encounter with the alien. In its own way, generativity allows the full incommensurability of emptiness, even if emptiness calls generativity into question.

3. The problem of dualism and its challenges

In my remarks, I have tried to explain by phenomenological means how we might be able to account for the project of an issue like "Phenomenology in Japan," and how the issues raised by the contributors to this special collection, issues such as Western dualism, could occupy the place it does in their writings and be troubling even for themselves. I also wanted to suggest a style of thinking capable of addressing the problems faced with "dualism," a style of thinking that seems peculiarly attractive to contemporary Japanese thinkers, namely, phenomenology.

Having already stated above that one problem consistently evoked in the essays that follow is the problem of dualism, I want to emphasize that while not all of the papers here make the problem of dualism their main theme, the issue does run through them like a guiding thread of thought, and at least allows them to be read through this particular focus, if not others. In what follows, therefore, I would like to summarize each of the authors' works, and by highlighting their main themes, evoke the problems of dualism and phenomenological clues to it. This will show at least one way in which phenomenology even if read against itself is presently being appropriated in Japan.

Toru Tani's cogent contribution, "Inquiry into the I, Disclosedness, and Self-Consciousness: Husserl, Heidegger, Nishida" formulates the problem of dualism in terms of a subject-object style of thinking. More specifically, in Husserl this style of thinking is assumed at first by understanding consciousness as an activity that constitutes the meaning of Being such that the gift of the world is not received, or at least is received only when consciousness is prepared to do so.

In contrast, as Tani lucidly points out, the phenomenon of passivity in Husserl proves to be more primordial than that of activity, and coincides with the recognition of the pregivenness or gift of the world, with pre-Being, such that consciousness is compelled to receive the gift even if it is not ready to do so. The recognition of primordial passivity in Husserl serves as a transition to Heidegger and his emphasis on the fundamental passivity of *Dasein*, who as *Da* or disclosedness of and openness to world, is a Can-Be and not always an I-Being. According to Tani, although Heidegger took pains to consider the pregivenness of Being and the structure of disclosedness, Heidegger did not pursue the latter far enough. For this reason he invokes the philosophy of Kitaro Nishida.

While there are some parallels between Husserl and Nishida, and even closer ones between Heidegger and Nishida, Tani elucidates the radicality of Nishida's thought by focusing on Nishida's "topos of absolute nothingness" as "self-consciousizing consciousness." By the latter expression, Nishida is rejecting a consciousness that is an object of the act of consciousness, and affirming the unity of consciousness with no separation, a spontaneous self-disclosing that is a united disclosedness of itself. Belonging properly to religious experience, this no-separation can overcome subject-object dualism because rather than polarizing, self-consciousizing-consciousness illuminates itself from within as the self-illumination of the topos.

In his "The Relationship between Nature and Spirit in Husserl's Phenomenology Revisited," Tetsuya Sakakibara locates the contemporary devastation of the environment and forms of interpersonal violence in a dualistic way of thinking and living that has dominated Modernity, and he draws on Husserl's lectures on "Nature and Spirit" to overcome this dualistic paradigm. Tracing in extraordinary and admirable detail the relationship between nature and spirit in Husserl's published and unpublished work, Sakakibara examines the motivations for spirit's "double forgetfulness" of nature, the dualism that is a product of this forgetfulness, and the implications it has for contemporary humanity.

In one respect, Sakakibara highlights Husserl's own phenomenological descriptions of nature and spirit and Husserl's own forgetfulness of nature as exemplary of our own forgetfulness. But in showing Husserl's recovery of this forgetfulness, Sakakibara suggests that this recovery is a possibility for us too. Ultimately, this forgetfulness is a double one: on the one hand, there is what might be commonly understood as a forgetfulness of the lifeworld, a forgetfulness of the lived-body and its environing-world, and a forgetfulness of the normality and abnormality of the lived-body through the quantification of nature peculiar to Modern spirit. On the other hand, and subtending this as the condition for its possibility, we have spirit's forgetfulness of its own natural basis, "passivity." The difficulty here is that in the very development of its natural basis and the spiritualization of that basis, spirit inevitably and necessarily "forgets" that nature "prior" to spiritual accomplishments. This primordial nature as the basis or side of spirit is always already forgotten.

If phenomenology is a constant effort to logify what inherently resists logification, and if Husserl at times fell victim to double forgetfulness of nature, reinforcing a kind of dualistic thinking, Sakakibara argues, it is nevertheless through Husserl's phenomenological analyses especially his *genetic* analyses of passivity that the double forgetfulness can become apparent to us and justifies our continued practice of phenomenology today.

Yet another mode of dualism, that of form and content, is investigated by Shigeto Nuki in his "The Theory of Association after Husserl: 'Form/Content' Dualism and the Phenomenological Way Out of It." In his writings characteristic of a "static" phenomenology, Husserl described the constitution of sense through the animating apprehension (*morphe*, form) of sense-data (*hyle*, content). But as Nuki indicates, it was his concrete phenomenology of passive genesis especially evident in his lectures on "passive synthesis" during the 1920s that pointed a way out of this kind of dualism.

But Nuki does more. Insightfully honing in on Husserl's third division of the *Analysen zur passiven Synthesis* and the phenomenon of affection, he shows how the phenomenon of affection and transmission of awakening, as the presuppositions of the constitution and unification of sense for an ego, allow us to see how material and formal syntheses are inextricably intertwined. Then, by drawing on some of Husserl's unpublished manuscripts, with certain allusions to Merleau-Ponty, and through his own enlightening examples, he responds to the question of the epistemological function of sensory hyle in perception that arises from dualism. This leads him to a critique of Derrida's reading of Husserl, highlighting not only the iterability of remembering, but the role of the "optimal" as the futural in-itself-for-us that is not accessible to a third-person perspective, but only to the point of view inherent in experience as it is taken up by the phenomenologist.

Another way that the dualist style of thinking gets played out is in terms of objectivistic accounts of experience that result in the dichotomous positions of realism and idealism, positions that are at the heart of the so-called mind-body problem. On this score, Junichi Murata presents us with a splendid essay using the phenomenology of color as its leading clue to work against the prejudices of realism and idealism (colors are either objective qualities, or colors are purely subjective appearances like hallucinations). In his "Colors in the Lifeworld" Murata gives us an incisive phenomenological account of how colors exist in experience, first, by taking the example of "mixed color" (e.g., the color of yellow in the television screen that is not a pure spectral color) and giving an account of it through Husserl's concept of perspectival adumbration. In order to describe the radical consequences of Husserl's notion of adumbration, Murata illustrates how color cannot be perceived independently from other situational contexts, and he introduces along with Merleau-Ponty (and Husserl) a phenomenological model of embodiment as well as the constitutive notion of normality as optimality (i.e., a kind of compossibility in and through incompossibility, in short, depth). After interpreting the ontological place of colors to be the whole lived-body/world nexus or "lifeworld," Murata is in a position to consider the coexistence of lifeworlds – e.g., between the lifeworld of bees and of humans – even though the colors of life other than human might be incommensurable with the perception of colors as they are given to humans.

In its own way, this evokes the problem of homeworlds and alienworlds mentioned above. Even though the constitutive notions of normality and abnormality for Husserl are circumscribed in its broadest parameters by a species, Husserl does account for the possibility of another species being a "homecompanion" or a normal "home-body" in "our" homeworld. Murata makes this case, implicitly, by noting that even though the colors of other animals (e.g., the ultraviolet colors "seen" in the lifeworld of bees) cannot be considered to be other aspects of *our* visible colors, the color of flowers that have co-developed with the perceptual organs of bees can be enjoyed by us from "our" perspectives, since they are aspects of the "invisible" color. In this regard, Murata's analysis proves to be another way of undercutting the dualism between spirit and nature.

Hiroshi Kojima's "On the Semantic Duplicity of the First Person Pronoun 'I' " is an original and creative phenomenological reflection on something that is taken for granted in everyday experience: the first person pronoun I. How is the I constituted? How is linguistic meaning possible? These questions are addressed by discerning the different dimensions of the I and their mediation through the second person pronoun, Thou. Kojima does this by describing the "serial I" as the self-consciousness of the physical-body [*Körper*] that exists and acts in the plural as reversible with and juxtaposed to other Is in homogenous space. This dimension is contrasted with the I that has no parallel to another, the absolutely unique, centrating I, the "primal I." The primal I is the consciousness of the lived-body [*Leib*] understood as my radical facticity as monad. It is the subject of imagination, temporally protracted in the abiding present. The (incomplete) unity of these two dimensions of the ego is the kinaesthetic I who inhabits a lifeworld and who is the lived-body passively pluralized.

I cannot hope to capture in this short space the subtleties of Kojima's analysis pertaining to the kinaesthetic ego as they concern mood and atmosphere, or his subsequent interpretation of the noema as the semantic focus of the optimally generalized image and its relation to the omni-temporal schema. Suffice it to say that after describing the duplicity of the lifeworld as the extensional perspective space and open field of image corresponding to the duplicity of the kinaesthetic ego who inhabits the lifeworld, as well as describing the possibility for the constitution of special communal lifeworlds through sensibility, emotion, sympathy and antipathy (what we have called above homeworlds and alienworlds), Kojima concentrates on what has been lacking thus far for an account of the constitution of the ego and the genesis of language.

Leaving the terrain of what could be called "Husserlian" up to this point, Kojima engages in a creative appropriation of Martin Buber's "I and Thou." And it is here that Kojima's analysis bears on the problem of dualism in the field of social ontology. According to Kojima's analysis, there is a complete coincidence of the I-Thou on both sides of the relation, reciprocally duplicated, such that the Thou does not only appear to me in a face to face encounter, but the same Thou occurs in myself, mediating the primal and serial I. This is not only a mediation that occurs between human beings with language, Kojima concludes, but between the artist and the natural world promoted by the Thou. It is in this way, writes Kojima, that we can understand Nishida's "self-identity of absolute contradictories."

Challenging the primordiality of dualities such as subject and object, or person and thing, Tadashi Ogawa gives a perceptive phenomenological description of a cultural invariant "Qi/ki" in his contribution, "Qi and Phenomenology of Wind." After providing the reader with a vivid historical orientation to the concept of qi/ki in the East as the living and moving, vital and spiritual power embodied in the human being, and the vital and spiritual power between heaven and earth, Ogawa develops a phenomenology of wind. Not only the East is aware of this phenomenon, Ogawa provocatively maintains, but the West as well, not only in such notions as the New Testament *pneuma* or the Ancient Greek *psychein/psyche*, but in the contemporary concepts developed in phenomenology of primordial passivity (Husserl), mood (Heidegger), and atmosphere (Schmitz).

Exploring three modes of mood or atmosphere as the "place" from which things, human bodies, the world, etc., are articulated, Ogawa shows *how* atmosphere/mood opens up the relation between the individual and world. First, whereas the philosophical tradition presupposes the separation of inner and outer worlds in human beings, atmosphere is prior to this separation; it comes over us and things radiate into it. Subjectivity, substantiality and its properties are broken down into the atmosphere of the surrounding thing since the atmosphere is the mode of the appearance of world and thing, and that through which the perceiving person is intertwined with the perceived world. Second, atmosphere or mood resists the fragmentation of sense-fields, since it is disclosed in a symphony of senses, in synaesthetic experience; I am pre-predicatively aware of atmosphere as mood in my synaesthetic experience. Finally, the relation of my body to the world is the very appearance of qi/ki/atmosphere that is the most profound and passive dimension of Being. Atmosphere as mood gathers and concentrates in me as lived-body.

Having oriented my precis of these essays along the lines of the problem of dualism and their challenges to it, let me conclude by stating what perhaps is already obvious to most. It is only thanks to the authors' masterful facility with Modern Western languages, in particular, with English, French, and German (to say nothing of their command of Classical languages like Greek and Latin) that we in the West are able to benefit from their original work in phenomenological philosophy. Since the majority of us cannot boast this kind of facility with Eastern languages like Japanese, it would remain a

238

difficult task for us to reciprocate this kind of generosity *in fact*, even if it *were* possible *in principle*. It is therefore our honor and our privilege to be able to present their work in this collection.[12]

Notes

1. I am indebted to A.R. Luther, *A Dialectics of Finite Existence: A Study of Nishida Kitaro's Buddhist Philosophy of Emptiness* (unpublished manuscript). See also Tadashi Ogawa, "Kitaro Nishida," in *Encyclopedia of Phenomenology*, ed., Embree, et al. (Boston: Kluwer, 1997), 490–494.
2. See the "Preface" to *Japanese and Western Phenomenology*, eds., Blosser, Shimomissé, Embree, and Kojima (Boston: Kluwer, 1993). See also Hiroshi Kojima, "Japan," in *Encyclopedia of Phenomenology*, ed., Embree, et al. (Boston: Kluwer, 1997), 367–371, and Hisashi Nasu "Sociology in Japan," in *Encyclopedia of Phenomenology*, ed., Embree, et al. (Boston: Kluwer, 1997), 655–659.
3. Ed., Yoshihiro Nitta (Freiburg: Alber, 1984).
4. Ed., Hiroshi Kojima, 1989.
5. See note 2 above.
6. *Human Studies*, eds., G. Psathas and K. Okunda, Vol. 15, No. 1 (1992).
7. See Tani's contribution in this collection, "Inquiry into the I, Disclosedness, and Self-Consciousness: Husserl, Heidegger, Nishida."
8. See Max Scheler, *Vom Ewigen im Menschen*, GW Vol. 5 (Bern: Francke, 1954), esp., 198–210.
9. For a more detailed exposition of the generative relation between home and alien, see my *Home and Beyond: Generative Phenomenology after Husserl* (Evanston: Northwestern University Press, 1995), esp. Section 4.
10. Regarding the liminality of experience, see Anthony Steinbock, "Limit Phenomena and the Liminality of Experience." *Alter: revue de phénoménologie*, No. 5 (1998).
11. Concerning the aforementioned "point," see *Luther, A Dialectics of Finite Existence*.
12. On a final note of thanks, I would like to express my gratitude to Bob Scharff for encouraging me in this editorial endeavor.

Continental Philosophy Review **31**: 239–253, 1998.
© 1998 *Kluwer Academic Publishers.*

Inquiry into the I, disclosedness, and self-consciousness: Husserl, Heidegger, Nishida

TORU TANI
Josai International University, 1 Gumyo, Togane-shi, Chiba, 283 Japan

> ... seeing the shape of the shapeless, hearing the
> voice of the voiceless ...
>
> Kitaro Nishida (IV, p. 6)

Introduction

Consciousness – *Bewußtsein* – was one of the key concepts of Husserl's phenomenology. In contrast to this, Heidegger – regarded as Husserl's most outstanding pupil – placed *Dasein* at the center of his own phenomenology. This change in key concepts may be seen as an upheaval in the phenomenology that purports to study the "things themselves": as a shift of focus from the activity of a *Bewußtsein* that constitutes the Being of objects, to the passivity of a *Dasein* that receives the donation of Being. But there is another aspect of *Dasein* that lies in the concept of *"Da"* (disclosedness), and it implies another possibility for the development of phenomenology that Heidegger did not fully develop. Kitaro Nishida, the Japanese philosopher who introduced phenomenology to Japan during the first half of the 20th century, developed this concept in his own way by analyzing the structure of disclosedness in terms of self-consciousness. In this paper I will follow the development of phenomenology from the "I" of Husserl, through Heidegger's concept of "disclosedness," to "self-consciousness" as it is understood by Nishida.

I. "Consciousness" according to Husserl

Husserl's concept of consciousness is often misunderstood. In a typical misunderstanding, one thinks of consciousness as something substantial. To avoid this misunderstanding, we must refer to the concepts of Franz Brentano inasmuch as they influenced Husserl.

Brentano distinguishes between "psychic phenomena" and "physical phenomena." This may remind us of the distinction made by Descartes between *res cogitans* and *res extensa*, when in fact the two pairs of concepts do not at all correspond. The distinction between psychic and physical phenomena is the distinction between the "act" of consciousness and the "object" of consciousness. Both psychic and physical phenomena are regarded, from the viewpoint of consciousness, as being distinguishable constituents of that consciousness. There is no act that is not directed toward an object. The act is always that which is directed toward a corresponding object. This directedness is called "intentionality," a concept to which I will return later. The act, the psychic phenomenon, contains the object, the physical phenomenon, inasmuch as the latter is internally experienced by the former. The latter exists in the former. Thus Brentano says: "Intentional inexistence [=internal existence] is characteristic of the psychic phenomenon. No physical phenomenon has a similar character, Therefore we can define the psychic phenomenon in the following way: The psychic phenomenon is a phenomenon that contains the object intentionally within itself"[1] This is an idea that Husserl adopted from Brentano.

There is another misunderstanding regarding consciousness. Consciousness as an act perceives the object. Does consciousness also perceive itself then? The correct answer is No. Here too Brentano avoids misunderstanding by distinguishing between "primary objects" and "secondary objects," although the word "object" is itself misleading. To avoid confusion, I will use the word "Object" in upper-case for the German word "*Gegenstand*," while using "object" for the German "*Objekt*." The Object of consciousness is a primary object, while the act of consciousness is a secondary object. For example, when we hear a sound, "we can call the sound itself a primary object, and the act of hearing itself a secondary object."[2] The sound itself is *thematically* a primary object of consciousness, while the act of consciousness is *athematically* a secondary object of consciousness. In any case, both are objects of consciousness. Husserl also received this idea from Brentano.

However, to avoid a misleading use of words, Husserl gives psychic phenomena another name: he refers to them as "intentional internal experiences [*intentionalen Erlebnissen*]." "We will thus avoid the expression psychic phenomenon altogether, and when accuracy is required, we will speak of *intentional internal experiences*" (Hua XIX/I p. 391), says Husserl. Intentional internal experience (thematically) has an Object as its primary object and (athematically) has itself as its secondary object. Husserl expresses this distinction by using the words "perceive" and "internally experience." Thus, the Object is perceived but not internally experienced. The internal experience itself is internally experienced but not perceived, and it is this experience that Husserl calls "consciousness." We can see this mode of thinking in the following sentence: "For example, in the case of outer perceptions, the sensation-moment of color, which is a *reell* [neither real nor intentional, but contained] constituent of the concrete act of seeing (i.e. in the phenomenological sense of visual perceptual appearance), is an 'internally experienced [*erlebt*]' or 'given-to-consciousness [*bewußt*] content,' just like the act of perceiving, and just like the perceptual appearance of the colored Object. On the other hand, the Object itself is perceived but neither internally experienced nor given-to-consciousness" (Hua XIX/ I p. 358). The word "consciousness" refers to the internal experience that is (athematically) internally experienced by itself, but not (thematically) perceived.

When Husserl speaks of consciousness in the following sentence, he is speaking of the internal experience of consciousness, and not of a perception of itself. "It is even nonsense to speak of an 'unconscious' [not given-to-consciousness] content which is afterwards given-to-consciousness for the first time. Consciousness [*Bewußtsein*] is necessarily a being-given-to-consciousness [*Bewußtsein*] in all of its phases." (Hua X p. 119) Here we should confirm that this sentence is not evidence of a metaphysics of presence in Husserl's thinking, if the word "presence" is taken to mean something thematic. He means rather that consciousness is internally experienced in an athematic way.

Consciousness is not punctually present or present as a point. Rather, it forms the breadth of the present. It possesses the internally experienced structure of retention–original-impression–protention. That is to say, it possesses not only a phase (appearance) given as original impression, but it also retains a phase (appearance) which is going away, and one that protends what is about to come. Husserl refers to the intentionality of retention and protention, which constitutes the unity of the Object, as a "crosswise intentionality [*Querintentionalität*]" (Hua X p. 82), and the retention in this case is called an "'outer' retention" (Hua X p. 118). Consciousness, in synthesizing the retentional phase (appearance), the original-impressional phase and the protentional phase, intentionally constitutes one Object. For example, in the case of the constitution of the squareness of a desk, consciousness constitutes a square by synthesizing the internally experienced retentional appearance of a parallelogram, the original-impressional appearance of a trapezoid, and the protentional appearance of a parallelogram. The square with right angles is perceived, not internally experienced. On the other hand, the appearances that do not have right angles are internally experienced, not perceived. (Cf. Hua XXI p. 282)

Consciousness also experiences *itself* internally; that is to say, it also experiences itself retentionally, original-impressionally and protentionally. The intentionality of the retention, original-impression, and protention of consciousness itself is referred to as a "lengthwise intentionality [*Längsintention-*

alität]" (Hua X p. 81), and the retention here is called an "'inner' retention" (Hua X p. 118). The phases of consciousness appear (as appearances) to consciousness; they are internally experienced. Consciousness constitutes the "I" by means of a "constant unity of coinciding with itself" that is similar to the way in which an Object is synthesized. The "I" is a self-consciousness which is constituted by the lengthwise intentionality of consciousness as the unity of an act. We know that Husserl did not accept such an I in the beginning, but that he later came to accept it clearly. He says: "In my old doctrine concerning the consciousness of internal time, I treated the intentionality shown here precisely as intentionality: as that which is protended as protention and modified as retention, but which remains a unity. However, I did not speak of the I, nor did I characterize intentionality as being an I-related intentionality (will-intentionality in the widest sense). Later I introduced such an I-related intentionality . . ." (Hua XV p. 5 94f.)

The later introduced I-related intentionality is qualified in the last half of this sentence as one that is "founded on an I-less intentionality ('passivity')" (ibid.). I will later consider this special "passivity" which Husserl places in quotes. But here, let me point out that the constituted I by means of which intentionality becomes I-related has two stages: In the first stage the I is implicit. The implicit I becomes explicit for the first time by means of "reflection." Reflection explicates only what is implicitly internally experienced or given-to-consciousness. Therefore, reflection is an explication of the I that is implicitly internally experienced (Cf. Hua III/I p. 123; Hua IV p. 101f.). On the one hand, the I contains in itself the I-pole, which has no determination of meaning but is only the unity of a constituting function somewhat analogous to an "S" without a "p" (to which I will refer later); on the other hand, the I also acquires a habituality that is analogous to the predicative meaning "p." Although we sometimes explicate and thematize such an I by means of reflection, we experience it internally before reflection, and are thus already conscious of it. In this sense, the I, as a unity of consciousness, is itself self-consciousness, although only athematically.

Now we know that there are three stages to the self-constitution of consciousness: the I-less stage, the implicit I, and the explicit I.

I have referred to Husserl's concept of constitution. But what does it mean? For one thing, it is radical. That is to say, constitution first makes possible the "meaning" of an Object; namely, it determines "what" the Object is. We can put it in the following way: If the Object is a brown and wooden desk, consciousness has "constituted" the meaning of "brown" and "wooden" from the internally experienced appearances of that desk. Even the meaning of "desk" has been "constituted" by consciousness. Without the operation of consciousness, no determination of the meaning of the Object is possible.

But the Object itself is individual. This desk is different from others. The meanings of "brown," "wooden," and even "desk," however, are general in the sense derived from "genus." There is no criterion for the distinction of the individual contained in meaning itself, which is general. The individuality of this desk is not determined by its meaning. It is determined by its "place in time." This latter "constitutes individuality for the first time" (Hua X p. 67). Thus, difference in meaning is not sufficient for the determination of individuality, whereas even where the meanings of two Objects are very similar or even identical, individuality is made possible by difference in place in time. This is a second aspect of Husserl's concept of constitution.

Moreover, "this desk" is (=exists) not in fantasy but in reality. But if I imagine a desk of gold, it is not in reality but in fantasy (I might say that it "would be" a desk of gold). This character of being a "real-Being" or an "imaginary-Being" is also "constituted" by consciousness. Being is also said to be "transcendent." In relation to the concept of "transcendent/Being," the consciousness which enables the determination of Being is said to be "transcendental." "In it [i.e. transcendental subjectivity] the Being of all things that the subject can experience in various ways, namely the transcendent in the widest sense, is itself constituted. Therefore it is called transcendental subjectivity."[3]

Husserl refines this conception of the constitution of Being. Being is constituted by making the Object belong to time, which is also constituted by consciousness. If the constituted time is "objective time," the Being of the Object is "real Being." If the constituted time is "quasi-time," then the Being of the Object is "imaginary Being." Time functions as the "form" which enables the determination of the Being of the Object: it is "time-form." Consciousness constitutes the Being of an Object by making it belong to this "time-form." What then can be said about space? Husserl regards space as being secondary in relation to time. Nevertheless, space also functions as a "form" which enables the determination of the Being of the Object. Husserl in his later years synthesized time and space into the "world," or better, "world-form." Then the Being of the Object came to mean the way in which the Object belongs to the world-form. For example, the city of Troy was for a long time considered only imaginary Being, but subsequent to the excavations of Schliemann, it reachieved real Being by acquiring a place in the world-form (in objective time and space). This operation of consciousness of making an Object belong to the world-form is called "positing." The constitution of an "imaginary Being" is a modification of "positing." This is the third aspect of Husserl's concept of constitution. Here, we can re-define the constitution of the individuality of an Object more precisely as the determination of its place in the world-form.

Consciousness constitutes the meaning, the individuality and the Being of the Object. This is the direct operation, or direct experience, of consciousness. And it is also the foundation of the judgment "S is p." The judgment "S is p" is a development of this direct constitution. Namely: S (directly experienced as individuality) is (directly experienced as real-Being) p (directly experienced as meaning).

I have spoken of the constitution of the Object. But how is the world that enables the constitution of the Object constituted? We must distinguish between Object(s) and the world, especially where Being is concerned. The Being of the world itself cannot be determined by its belonging to the world-form. This is because the world as the framework that contains Objects cannot be contained within that same framework. Nonetheless, only that which is posited – that is, made to belong to that framework – has "Being." Does the world then not "exist"? It exists, but its way of Being is different from the Being of the Object. It exists without and before positing. *"The consciousness of the world is consciousness in the mode [of Being] of certitude of belief. It is not acquired by an act of positing of Being that enters particularly into the connection of life [Lebenszusammenhang] – by an act of grasping as Being-there [daseiend] – nor moreover by a predicative judgment of existence. These [=positing and the predicative judgment of existence] already presuppose the consciousness of the world in the certitude of belief."* (EU p. 25) The world exists before positing and in a different way from the Object. Husserl has changed his old conception of the world as a "sum-total of all Objects" (Hua III/I p. 11).

The world as the sum-total of all Objects is the product of an active and higher constitution. It presupposes a passive and lower constitution. With regard to time, at the level of passive and lower constitution, retention and protention function together with the original impression, although the word "function" at this passive level means no more than "keeping." But Husserl goes back to the most passive level: to primal-passivity [*Urpassivität*]. It is at this lowest level that the world is first given. The Being of the world at this level precedes Being which is constituted by the operation of consciousness. It is "pre-Being." The pre-Being of the world is not a product of the constitution of consciousness at all, but a gift of the world itself. Consciousness only receives it, or better, consciousness is compelled to receive it even when consciousness is not yet prepared to do so. The more passive consciousness becomes, the more the world gives itself.

In this primal passivity (that is to say, in the "passivity" with quotes which signifies the special, original passivity) intentionality does not sufficiently operate yet. It is a level of minimalized intentionality, or "pre-intentionality."[4] Husserl's inquiry into pre-intentionality began in the 1930's, but the pre-stage for that inquiry can be found in the concept of "*Urbewußtsein*" in his *Phenomenology of the Consciousness of Internal Time* (cf. Hua X p. 118f.). Iso Kern has related this concept to Buddhist thought.[5] Lengthwise intentionality does not yet sufficiently operate. The constant unity of coinciding of the I has not yet been achieved. No explicit I, nor even an implicit I, has yet been constituted. Here we can recall the sentence: The I-related intentionality is "founded on an I-less intentionality ('passivity')." We have arrived at the most original level of the self-constitution of consciousness.

Here we can ask if there is no unity of consciousness at this level. Husserl finds a minimal unity that does not depend upon the operation of consciousness. It is a delicate situation that is difficult to characterize because our ordinary language has a tendency to substantialize. Although this minimal unity is not yet the "I" in the proper sense, Husserl refers to it as the I, saying: "the constant I [is] the constant original source [of constitution], it is identical not by an 'identifying' but as an originally united being, it is in the most original pre-Being."[6] The unity is not established or substantialized; it is very fragile and transitional. It is a minimal unity. This unity means, from the point of view of the consciousness, that there is already a minimal self-consciousness, although Hussert does not state this point clearly.

II. The Disclosedness of *Dasein* according to Heidegger

Heidegger was influenced by Husserl's phenomenology and by Dilthey's philosophy of life. His starting point was the concept of *Dasein* (Being-there), where the influence of Husserl and Dilthey can be seen.

Heidegger introduced the concept of "life = *Dasein*" in his early years (HGA 61 p. 85). This concept implies the "facticity" (ibid. p. 76) of "historical" life. Each *Dasein* is thrown into the fact of history. There is no necessity and no choice. *Dasein* must assume its facticity passively.

On the other hand, the concept of *Dasein* also relates to the concept of consciousness. In German, consciousness is "*Bewußtsein*," thus sharing a "*sein*" with "*Dasein*." Heidegger explains this in the following way: "The expression 'There (*Da*)' means this essential *disclosedness* [*Erschlossenheit*]. By this disclosedness this being [*Dasein*] exists for itself 'there' with the *Dasein* of the world... *Dasein is its own disclosedness*."[7] Here we should recall the original German meaning of consciousness or *Bewußtsein*. "*Bewußt*" means almost the same thing as "known." Thus, "Bewußtsein" is equal to "Being known." "Known" can mean not only "thematically known" but also "athematically known." According to Husserl, "*Bewußtsein*" is equal to internal experience, which is athematic. Therefore, consciousness in Husserl's case means an athematic Being-known of the world and of itself. This conception of Husserl's corresponds to that of *Dasein* as Disclosedness in Heidegger's thinking.

Yet Heidegger chooses to introduce the concept of Dasein instead of speaking of *Bewußtsein*. Why? There is an implicit critique here of Husserl's concept of constitution. The critique has two aspects. On the one hand, if consciousness constitutes Being (of the Object), the former is original while the latter is derivative. But for Heidegger, Being is always original. It is given (in passivity) before the active constitution of the consciousness. The idea of a constituting consciousness, which pretends its own originality, is unsuitable if we are to approach Being as original. Husserl's concept of constitution is too strong for Heidegger. Heidegger himself wants to explore the original givenness, or donation, of Being.[8] However, if consciousness constitutes (not only the Object but also) itself as the I by lengthwise intentionality – or so Husserl thought when Heidegger was his pupil, although he, Husserl, encountered the I-less passivity in his later years – then consciousness [*Bewußtsein*] must always mean an I-Being [*Ich-sei*]. But for Heidegger, *Dasein* is not always an I-Being. The self of *Dasein* can also be an "anonymous one-self [*Man-Selbst*]." There is always the possibility of a "deterioration." Thus, the Being of *Dasein* is not an I-Being, but a Can-Be. It can be an I (in all its authenticity), and it can also be an anonymous one-self (in deterioration). Husserl's concept of constitution and therefore of consciousness is too strong and too active for Heidegger. For this reason, they must be replaced by *Dasein* and the concept of Disclosedness.

In order to understand the most original experience totally, the close connection between the givenness of Being and the structure of Disclosedness must be thoroughly explored. Although Heidegger considered the former with great care,[9] he did not pursue the structure of Disclosedness any further. For this reason, I will move on to the work of Kitaro Nishida, who approached the problem of original experience mainly in terms of self-consciousness.

III. Nishida and the self-consciousizing-consciousness

Kitaro Nishida is regarded as one of the representative philosophers of modern Japan, and often as the only one to have constructed a philosophical system. His philosophy is difficult to understand for various reasons, many of which stem from the delicate and sometimes clumsy maneuvering between traditional Japanese/Oriental modes of thought and newly encountered Western ones. Many contemporary philosophers have attempted to interpret him. I will attempt another interpretation by approaching his philosophy through phenomenology on the one hand, and by trying to find a new possibility for phenomenological thinking in his distinctive ideas, on the other. My attempt is directed towards a *dia-logos* between Nishida and phenomenology.

Husserl inquired into the conditions for the possibility of the judgment "S is p." The predicate "p" is "general" in principle. Even "species" are general and not individual, although they are less general than "genera." That which determines individuality is a place in time (and space), or a place in the world-form.

Nishida agrees that the predicate does not determine individuality. But he also speaks of a final individual "S" that lies beyond genera and species. With this sense of "beyond" in mind, he names it the "transcendent subject," which when universalized becomes the "judgmental universal." However, even the transcendent subject must be recognized concretely as such. This means that there is something that subsumes the transcendent subject, although it is beyond such normal predicates as "p" in the sense of genus or species. Nishida calls that which subsumes it the "transcendent predicative plane." This a famous concept of Nishida's, but what is it precisely? Is it something like a speculative postulate which is only thought?

Let us interpret the concept of the "transcendent predicative plane" from a phenomenological point of view. Phenomenologically speaking, that which determines individuality beyond genera and even beyond species is a place in time and space. Even an individual subject which is a transcendent subject must belong to time-form and space-form to the extent that it possesses individuality, and thus it must belong to world-form, which is different from and therefore beyond the determination of the meaning "p." In this way, Nishida can be interpreted phenomenologically instead of speculatively. He says: "The judgmental universal means that which subsumes what is not predicate but only subject, i.e. what is individual. It is to be regarded as a so-called concrete concept because it contains within itself the principle of individualization. The predicative plane of the universal, i.e. its topos, is to be regarded as the so-called objective world." (Nishida V, p. 99; cf p. 61.)

But there is something further beyond the transcendent predicative plane, that is, the objective world, and thus the world-form, which subsumes them. It is called the "plane of consciousness." "What we regard as the plane of consciousness is something which transcends even the [judgmental] universal and which contains the latter in itself" (ibid.) The transcendent predicative plane, i.e. the world-form, is given-to-consciousness to the end; it is not independent of consciousness. This conception is similar to that of Husserl. For Husserl also, the world-form does not transcend consciousness. Although the original donation of the world-form is independent of consciousness, the range of disclosedness, or givenness-to-consciousness, of the world-form does not transcend consciousness. And although the world-form later expands, the expansion owes itself to the operation of consciousness.

The relationship between the plane of consciousness and the I is similar to that between the transcendent predicative plane and the transcendent subject. "The plane of self-consciousness is the topos where what we call the 'I' lies." (ibid.) That is to say, the plane of consciousness is co-extensive with "the universal-which-is-conscious-of-itself" that subsumes the I.

Up to this point, we can see some parallels between Husserl and Nishida. However, Nishida goes further to consider something that transcends even the plane of consciousness as "the universal-which-is-conscious-of-itself": He calls it "the intellectual universal." It is also called "the topos of true nothingness" or "the topos of absolute nothingness." What is this? In order to interpret it, let us paraphrase it in another way: It is not "consciousness which is consciousized," but a "self-consciousizing consciousness" – or so I will translate. For although the English language possesses only a noun form ("consciousness") and an adjective form ("conscious") for the matter at hand, Japanese also has a verb form which Nishida uses liberally. So for the sake of precision, I will introduce the word "consciousize," instead of using the more usual term "be conscious of." Thus, instead of speaking of "the universal-which-is-conscious-of-itself," we can now call it "the self-consciousizing universal."

Nishida rejects the idea of a "consciousized" consciousness – that is to say, a consciousness which is an Object of the act of consciousness. This means that he rejects the separation of the consciousizer and the consciousized. This separation is not apparent in the English terms "conscious" and "consciousness" (nor in the German "bewußt" and "Bewußtsein"), although the separation is clearly there. For a "conscious man" is always a "consciousizing man" and never a "consciousized man," and "consciousness" is always a "consciousizing" and never a "consciousizedness." The ambiguity of the word "consciousness" makes us blind to the separation, but in the case of self-consciousness, the separation becomes a crucial problem. The "consciousizing" consciousness is itself "consciousized." It is this form of "self-consciousness," where the consciousizing consciousness is separated from the consciousized consciousness, that Nishida wishes to reject.

Nishida speaks of the unity of consciousness that has no separation: This unity is what he wants to express by the term "self-consciousizing consciousness." Consciousness is most properly a united disclosedness of itself. In contrast to this, the plane of consciousness is only one separated – or better, "consciousized" – aspect of this unity. What Nishida calls "the intellectual universal" contains not only the consciousized plane, which contains the I as a so-called cognizant Subject, and the predicative plane, which contains the individual as a so-called Object, but it is also its own disclosedness. Nishida says of this intellectual universal that it "must be considered not only as a cognizant Subject which constitutes the world of Objects for the consciousizing Self, but also as an intuitive Self which, eliminating all positions, sees the inside of itself. However, this must not be a so-called consciousized consciousness, but a self-consciousizing consciousness" (Ibid. p. 149). With this concept, Nishida denies the separation of Subject and Object.

Already in his early years, Nishida introduced the concept of "pure experience" where "there is not yet Subject, nor Object, and knowledge and its Object are completely united" (Nishida I p. 9). He maintained and developed this concept, and later added the concept of an "intuition" that is intellectual. "Intuition is consciousness in constant movement, not yet differentiated into Subject and Object, a unity of the knower and the known, reality as it is" (Nishida II p. 15). The self-consciousizing-consciousness is a radicalization of these concepts, developed particularly in the direction of disclosedness. This concept is different from and goes beyond the Husserlian concept of reflection. Although Husserl attempted to assure the identity between the reflecting I and reflected I, the difference between them remained to the end. "The pole which is reflected in reflection is not the living pole . . . "[10] Why was this so? This can be understood by focusing on the concept of intentionality. Namely, Husserl considers the I to be differentiated by lengthwise intentionality and the Object by crosswise intentionality. Intentionality is divided into two poles. In the case of reflection, where intentionality thematizes itself, it is again polarized into the reflecting I and the reflected I. In this sense, intentionality/reflection is necessarily a differentiation or separation. In contrast to this, Nishida's self-consciousizing-consciousness lies not in the direction of intentionality as with Husserl, but of disclosedness, as with Heidegger (although Heidegger did not pursue the idea as far as Nishida). And because intentionality – which can be illustrated as having the form of a vector – does not operate here, consciousness as disclosedness consciousizes itself not in vector form, but in the form of a topos. Instead of polarizing, self-consciousizing-consciousness illuminates itself from within – or better, it is self-illumination of the topos – because it contains the plane of consciousness as well as the world-form. And it, as the topos, contains the two poles in the sense that such a polarization can take place only within it and because the separation can also be surmounted only within it. Because it contains the polarized two, it itself has no separation.

This conception is similar to Michel Henry's concept of "immanence" or "auto-affection of the act in immanence."[11] But in Nishida's case, consciousness is the disclosedness not only of itself but also of the all, since it contains the plane of consciousness which contains the transcendent predicative plane which contains all individuals. Thus we can understand Nishida's consciousness to be the disclosedness of the all. As he says: "Truly objective knowledge can be said to be the self-consciousness of the world itself" (Nishida X p. 397).

But we can still ask insistently: Is the unity of self-consciousness in no way disrupted? Nishida would probably answer in the following way: An absolutely unified self-consciousness is possible, but it belongs to the realm of religion. For philosophy, some disruption will remain. This is what makes possible the dialectical development of knowledge. But the origin and goal of that knowledge lies in unified self-consciousness.

This conception is similar to that of Merleau-Ponty, who was inspired not only by Husserl and Heidegger but also by Schelling. Merleau-Ponty speaks of an "'intellectual intuition,' which is not an occult faculty, but perception itself before it is reduced to ideas, perception sleeping in itself, where all things are me because I am not yet the subject of reflection."[12] Merleau-Ponty, who adopts the vantage point not of religion but of philosophy, and who therefore acknowledges the (minimal) disruptions within this intuition, also calls it "cognition by sentiment,"[13] and places it in a dialectical relationship with reflection.[14] Here also, the discrepancy between intuition and reflection motivates the dialectical development of knowledge, which is a realization of the original intuition, or of disclosedness. Nishida says: "In self-consciousness, when the self makes an Object of its own act and reflects upon it, the reflection is immediately an act of self-development or self-evolution. The development/evolution goes on thus infinitely." (Nishida II p. 15) Nishida and Merleau-Ponty are similar in this aspect.[15]

As divisions and separations are further eliminated, consciousness is further disclosed as self-consciousness. According to Nishida, this disclosure takes place spontaneously. There is a tendency so to speak to "self-disclose," although the tendency is weak and easily concealed by the separatist way of thinking. This conception of a spontaneous self-disclosing corresponds to Nishida's conception of "action-intuition," which was influenced by Fichte's "Tathandlung," where the absolute I is regarded as a unity of action [Handlung] and the fact [Tat] constituted by that action. But for Nishida, Fichte's Tathandlung is not radical enough. Fichte's absolute I is still objectified, and Nishida insists that it must be radicalized into the self-consciousizing consciousness.

Finally, we can try to understand Nishida's concept of self-consciousizing consciousness in contrast to Husserl's conception of Being. Husserl regarded the world-form as the condition of possibility for the determination of the Being of the Object. But the world-form itself has pre-Being. Therefore, both the Object and the world-form belong to Being in the widest sense. If we consider

Husserl's world-form and the Objects that have Being in that world-form as being analogous to Nishida's transcendent predicative plane and the individuals contained in it, we must say that the self-consciousizing consciousness is distinct from both of them. The self-consciousizing-consciousness itself has no Being, and is no Being. We can say that it is "Nothing." We can say almost the same thing in contrasting Nishida's conception to those of Heidegger and Merleau-Ponty. Although Heidegger also spoke of "nothing," in his case it was another name for Being. He opened the way to self-consciousness as disclosedness, but he did not develop this inspiration with the same radicality as Nishida, who arrives at an absolute nothingness.

In this way, by interpreting Nishida phenomenologically instead of speculatively, we encounter a possibility for the development of phenomenology that was not pursued by either Husserl or Heidegger.

IV. Dia-logos

If we follow the path of Husserl's thinking, constitution by consciousness was at first very active. Consciousness constitutes even the Being of the Object. Husserl's conception was very "aggressive" and "humanistic." But Husserl eventually encountered the donation of the world. When he accepted the idea of donation, he became ironically "quietist" in the double sense that, on one hand, consciousness came to be regarded as primally passive, and on the other hand, the donor of the most original world-form is suggested to be God (Hua XV p. 385). In Heidegger's case, the starting point was *Dasein*, which is not so active, and the donation of Being (of Nature) was approached in a more "naturalistic" manner. Both Husserl and Heidegger regarded this donation as taking place at the most passive level of consciousness. This development of phenomenology can be regarded as a radical criticism of the modern sciences that neglect the idea of donation altogether. But neither Husserl nor Heidegger went as far as Nishida did in his development of the theory of self-consciousness, where self-consciousness is seen as a spontaneous disclosing. In Nishida's philosophy, neither Being nor Nature are donors; they are entirely neutral and indifferent to any such donation. We are reminded of the words of Lao-tze: "Heaven and earth are not charitable."[16]

I have attempted to present the distinctiveness of Kitaro Nishida's thinking in relation to that of Husserl and Heidegger. The greatest difference lies in the absolute radicality of Nishida's conception of nothingness, which is motivated by and further radicalized by Nishida's ultimate (and perhaps intuitive) refusal to recognize the separation of Subject and Object. This difference, which is a fundamental difference of logos, is what makes Nishida both interesting and obscure to the modern reader, even to the modern Japanese reader. For somewhere in the decades between Nishida and ourselves, we (the Japanese) have become so permeated with Western-style Subject-Object thinking that it has come to be our starting point also, even when the purpose is to criticize it. We, or at least I, can no longer intuitively identify with Nishida's way of thinking. Thus the need to re-approach him by the pathways of phenomenology.

My purpose has not been to highlight the differences between Husserl, Heidegger and Nishida. The "things themselves" – which all three ultimately pursue, although by different paths – are not totally different, even if not completely identical. The "things themselves" lie between difference and identity. The delicate balance between difference and identity presses us to investigate what it is, that is, if not identical, not simply different. The *logoi* are also different; this is to say that a simple dialogue that presupposes an identity of grammar and vocabulary is impossible. But a *dia-logos*, as a confrontation of different *logoi*, can generate new investigations. And we, a new generation standing at a different vantage point, can bring their *logoi* into *dia-logos*, even after the death of the three thinkers. The fertility of phenomenology lies in the possibility of such a generative *dia-logos*.

Citations from the central texts of Husserl, Heidegger and Nishida are included in the main text, and refer to the following volumes:

Hua: Edmund Husserl: *Husserliana*, The Hague: Martinus Nijhoff, Dordrecht/Boston/London:Kluwer Academic Publishers.

HGA: Martin Heidegger: *Gesamtausgabe*, Frankfurt am Main: Vittorio Klostermann.

Nishida: *Nishida Kitaro Zenshu*, Tokyo: Iwanami-shoten.

Notes

1. Franz Brentano, *Psychologie vom empirischen Standpunkt*, Volume 1, p. 125f.
2. Brentano: *Ibid.*, p. 180.
3. Ms. M 1113, p. 2, cit. from Alwin Diemer, *Edmund Husserl - Versuch einer systematischen Darstellung seiner Philosophie*, 2. verbesserte Auflage, Meisenheim am Glan: Anton Hain 1965, p. 15.
4. Ms. C 17 IV p. 6 [1932], cit. from Klaus Held, *Lebendige Gegenwart*, The Hague: Martinus Nijhoff, 1966, p. 102.
5. Iso Kern, "Object, Objective Phenomenon and Objectivating Act according to the 'Vijnaptimatratasiddhi'of Xuanzang" (600 – 664).
6. Ms. A V 5, p. 5 [1933], cit. from Held: *Ibid.* p. 106.
7. Martin Heidegger, *Sein und Zeit*, Tübingen: Max Niemeyer 1977, pp. 132 – 133.
8. Heidegger is critical of all active operations of consciousness which constitute the Object and its Being thematically. Concepts such as "*Sorge*," which operate athematically and in a non-object-oriented manner, are closely related to this aspect of his critique. But such

concepts seem to me to be secondary in his critique of Husserl. The primary target of criticism is Husserl's idea of the constitution of Being.

9. To this end, he considers not only the matter of "*Geschick*" in the problematics of historicality, but also the ancient Greek concept of *physis* that reigns over all movement.
10. Ms. A V 5 p. 2, cit. from Held: *Ibid.* p. 121.
11. Cf. Yorihiro Yamagata, "The Philosophy of Nishida and French Philosophy" in *Nishida-Tetsugaku wo manabu Hito no tameni, Sekaishiso-sha*, 1996, p. 86.
12. Maurice Merleau-Ponty, *Résumé de cours*, Paris: Gallimard 1968, p. 107.
13. Maurice Merleau-Ponty, *Le visible et l'invisible*, Paris: Gallimard 1964, p. 302.
14. Merleau-Ponty, *La nature*, Paris: Edition de Seuil 1995, p. 73.
15. Their "dialectical" path as a way of recovering the original situation contains another problem: it does not encounter the radical Other – a problem I cannot investigate within the scope of this paper.
16. Lao-tze: 5th Stanza.

Note about the author

Toru Tani (born in Ichinomiya, Japan, 1954) was educated in philosophy at Keio University (Tokyo). His primary interest is contemporary Western philosophy with a strong emphasis on Husserlian phenomenology. In recent years he has concentrated on an analysis of what is given naturally to the primally passive consciousness and on what does not give itself, although it is encountered (the Other). This subject is dealt with extensively in a forthcoming book: *The Physis of Consciousness*.

Continental Philosophy Review **31**: 255–272, 1998.
© 1998 *Kluwer Academic Publishers.*

The relationship between nature and spirit in Husserl's phenomenology revisited

TETSUYA SAKAKIBARA
Department of Philosophy, Faculty of Letters, Ritsumeikan University, 56-1 Kitamachi, Toji-in, Kita-ku, Kyoto 603-8577, Japan

I. Introduction

The problem of the destruction of nature and the natural environment is one of the most serious problems that confronts humanity at the end of the twentieth century. This destruction has its basis in the dualistic way of thinking that has dominated the Modern age: Humanity has been accustomed to think of the world dualistically as a confrontation between subject and object, soul and body, and spirit and nature. Both Modern natural sciences and scientific technology in general are grounded in such a dualistic way of thinking. But this way of thinking itself, in other words, the Modern dualistic paradigm, is now one of the most serious problems of present-day philosophy. It has become clear that a new relation between humanity and nature must be established. In this paper, I will re-examine the relationship between "nature" and "spirit" in the context of Husserl's phenomenology in which, in my opinion, a new concept of nature can be found. In the course of this investigation, I shall try to elucidate the present-day situation of human spirit in a phenomenological way.

According to Marly Biemel, the editor of *Ideen II*,[1] Edmund Husserl (1859–1938) first began to deal with the problem concerning the constitution of various objectivities just after the pencil manuscript of *Ideen I* from 1912 in which he formulated and systematically described his phenomenological method for the first time. These phenomenological analyses of constitution, especially those concerning nature and spirit which were left as posthumous manuscripts (1912–1928), are now partly published in *Ideen II* (1952). In order to investigate Husserl's idea of the relationship between nature and spirit adequately, it will therefore be necessary to refer to those texts of *Ideen II*, while at the same time also turning to some yet unpublished, however highly relevant manuscripts.[2]

In what follows, I will first survey Husserl's analysis of the relationship between the naturalistic and personalistic attitude and, at the same time, of the relationship between nature and spirit in the original manuscripts of *Ideen II* (1912–1913)[3] (II). I will then go into his idea of the "natural basis of spirit" found in a supplement of *Ideen II*, examining a complex relationship between nature and spirit evident there (III). After sketching the development of this idea of "natural basis" briefly (IV), I will develop the theme of spirit's double forgetfulness of nature to be found in Husserl's texts and suggest its motivation (V). Finally, I will speculate on the possible meaning of phenomenologizing for contemporary humanity from this viewpoint (VI).

II. Naturalistic and personalistic attitude

In Husserl's phenomenological analyses concerning the constitution of nature and spirit (or person) found in the original manuscripts of *Ideen II*, the following two notions play the most important role; the naturalistic and the personalistic attitude.

According to Husserl, the "naturalistic attitude" is the attitude in which "nature comes to givenness and to theoretical cognition as physical, bodily, and psychic nature" ("H" 37 [M III 1 I 4, 69]: Hua IV, 208). In this attitude, not only a material thing but also "a human being and a human soul" are taken "as nature" (cf. "H" 12 [M III 1 I 4, 15]: Hua IV, 180f.). All these entities are ordered in "the one 'objective' world with the one space and the one time" ("H" 37 [M III 1 I 4, 69]: Hua IV, 209). Each of them is regarded as "a fact" in the "substantial-causal nexus of nature" ("H" 12 [M III 1 I 4, 15]: Hua IV, 181). He writes: "All the facts concerning the person," which occur in a unity with a body, appear also "as facts of nature" ("H" 16 [M III 1 I 4, 19]: Hua IV, 184). This is precisely what one considers as "natural-scientific" attitude ("H" 37 [M III 1 I 4, 69]: Hua IV, 208), in which "everything personal" is "subordinate" to "nature" ("H" 16 [M III 1 I 4, 19]: Hua IV, 185).

However, according to Husserl, our everyday life is not and cannot be performed in such an attitude. I have my "surrounding world [*Umwelt*]" in everyday life, and each human being found in my world appears as "a person who represents, feels, evaluates, strives, and acts and who, in every personal act, stands in relation to something, to objects in his surrounding world" (cf. "H" 17 [M III 1 I 4, 20] and 18 [M III 1 I 4, 25]: Hua IV, 185f.). I as a subject also appear as a person in my surrounding world. We, as persons, "live with one another, shake hands with one another, or are related to one another in love and aversion, in disposition and action" ("H" 14 [M III 1 I 4, 17]: Hua IV, 183). In everyday life, things in the surrounding world are not "mere things," but "use-Objects (clothes, utensils, guns, tools), works of art, literary products, instruments for religious and judicial activities [. . .]" ("H" 13 [M III 1 I 4, 16]: Hua IV, 182). This attitude of consciousness which is entirely different from the naturalistic one is now termed "personalistic attitude" ("H" 11 [M III 1 I 4, 14]: Hua IV, 180; "H" 14 [M III 1 I 4, 17]: Hua IV, 183).

In the course of his phenomenological analysis concerning the personalistic attitude, however, Husserl makes it clear not only that the personalistic and the naturalistic attitudes – and correlatively the region of spirit and that of nature – are entirely different, but also that objective, natural-scientific nature is nothing other than "an objectivity constituted in the context of the personal world" ("H" 38 [M III 1 I 4, 70]: Hua IV, 209). This objectivity presupposes personal intersubjectivity, namely, the nexus of persons. According to him, the "'objective' nature" in natural science is precisely what persons or spirits, who have their own surrounding world, have constituted in order to overcome "the differences in 'world pictures,' which come to the fore within intersubjective consensus" ("H" 36 [M III 1 I 4, 68]: Hua IV, 208). And this constituted 'objective' nature "henceforth belongs for its part to the surrounding world of the communal spirit [Gemeinschaftsgeist]" ("H" 36 [M III 1 I 4, 68]: Hua IV, 208). Husserl states clearly: "the naturalistic attitude is subordinated to the personalistic" ("H" 15 [M III 1 I 4, 18]: Hua IV, 183), only that one forgets this when one is working on natural sciences. The naturalistic attitude is only possible by means of "a kind of self-forgetfulness of the personal Ego [eine Art Selbstvergessenheit des personalen Ich]," the person's forgetfulness of "its surrounding world" ("H" 15 [M III 1 I 4, 18]: Hua IV, 183f.).

It is now clear that Husserl here, throughout his phenomenological analyses concerning these two attitudes of consciousness, maintains not only the dualistic confrontation between nature and spirit, but also a superiority of spirit over nature. If one also takes into account the last section 64 of the published version of Ideen II whose title is "Relativity of nature, absoluteness of spirit,"[4] it seems at first that the superiority of spirit remains unchanged in Husserl's late phenomenology.

However, is it true that Husserl remained trapped in this Modern dualistic way of thinking for the rest of his life? Did he really claim that spirit is "absolutely" superior to nature? I think this is not true at all. If one not only carefully investigates the main texts but also the supplements of Ideen II, it will become clear that Husserl, after the original manuscripts of Ideen II, came to speak of nature in another sense, and if this is taken into consideration, one can no longer speak of the one-sided superiority of spirit over nature. In order to show this, I will delve into a supplement of Ideen II, a phenomenological analysis concerning the psychic basis of spirit.

III. The natural basis of spirit

The text that I will discuss now is the first part of the supplement XII, entitled "Person – Spirit and its psychic basis (concerning the Ego as person)"
(Hua IV, 332–340). According to M. Biemel, the text was written at "the
end of January 1917" (Hua IV, 418). Husserl here characterizes the psychic
basis [*seelischer Untergrund*], the natural side [*Naturseite*], or the underlying
basis in nature [*Naturuntergrund*] of spirit phenomenologically. I will first
summarize his descriptions.

The Ego as spirit performs various Ego-acts, carries out various active
intentionalities. Before each of them, however, the "sensuous data" (Hua IV,
334), the "data of sensation" (Hua IV, 337), or the "sensuous impressions"
(Hua IV, 337) are pregiven to the Ego, and then the "sensuous feelings" and
the "sensuous data of instincts [*sinnliche Triebdaten*]" can occur as grounded
in data of sensation (Hua IV, 334). In other words, "instincts [*Triebe*] or tendencies" which intend an "arrival" of similar impressions or reproductions
originate from the given data of sensation (Hua IV, 337). Husserl calls this
level of sensibility "primal" (Hua IV, 334f.: "*Ursinnlichkeit*"). He also states
that the data and tendencies of primal sensibility are subject not only to the
laws of time-consciousness like that of "retention" (Hua IV, 335) or those
related to "protentions" (Hua IV, 337), but also to the "laws of association and
reproduction" (Hua IV, 336, 337. Cf. also 338). Those tendencies or instincts
of primal sensibility "move from the sensuous to the sensuous" (Hua IV, 337)
without any active participation of the Ego. They are "natural tendencies" (cf.
Hua IV, 339: "*Naturtendenz*") which "pertain to the sensuous itself" (Hua IV,
337). And all the "formations [*Gebilde*]" of "primal sensibility" sink down
naturally within the Ego – following the laws of time-consciousness – and
"pertain to the medium of the Ego [= the medium of its 'history'], to the Ego's
actual and potential possessions [*Habe*]" (Hua IV, 334). Since this sedimentation occurs without any participation of the spiritual Ego as "reason" (Hua IV,
334), Husserl calls those possessions of the Ego "secondary sensibilities not
originating from reason" (Hua IV, 334). According to him, these sensibilities
have "aftereffects in tendencies" or instincts (Hua IV, 338). He now terms
this level or stratum of sensibility – namely, that of the primal sensibilities
and their historical sedimentations not originating from reason – "authentic
sensibility" or "spiritless sensibility" (Hua IV, 334). It is on the basis of this
given data of authentic sensibility that "affections" directed to the Ego-subject
can occur, affections that motivate Ego's active intentionalities (cf. Hua IV,
337–339).

Since the authentic sensibilities and the affections founded in them occur
passively without any active participation of the spiritual Ego, i.e., before
all its intentional acts, they certainly belong to the "sphere of nature" (cf.

Hua IV, 338), and yet to the "natural basis of the soul [*Naturuntergrund der Seele*]" (Hua IV, 339) as the " 'psychic' basis of the spirit [*'seelischer' Untergrund des Geistes)*]" (Hua IV, 334. Cf. also 332, 333). On the other hand, the spiritual Ego, facing affections directed to it, can simply experience or accept the "stimuli" (= "receptivity" as "the lowest Ego-spontaneity or Ego-activity") (cf. Hua IV, 335), or it can furthermore perform its active, rational intentional acts on the basis of those affections. But following the laws of time-consciousness, all these rational Ego-acts and their formations also sink down to the sphere of sensibility and come to pertain to the medium of the Ego's history (cf. Hua IV, 338). These ever new sedimentations (which Husserl calls "secondary sensibility, which arises through a production of reason," "reason which has been degraded into sensibility," "unauthentic sensibility," or "intellective or spiritual sensibility,")[5] become pregivennesses for the Ego as well, pregivennesses that have after-effects in associative tendencies or instincts,[6] and which affect the Ego. Since this process of sedimentation also occurs passively without any active participation of the Ego, and yet within the sphere of sensibility, it also belongs to the natural basis of the soul. Thus, Husserl calls the dimension of sensibility, composed both of the authentic, spiritless and of the unauthentic, spiritual, the "psychic basis" (Hua IV, 332, 333, 334), "natural basis" (Hua IV, 338, 339), or also "natural side" (Hua IV, 338, 339) of the Ego as spirit.

It is now clear that nature as the basis of spirit as sketched above is not at all the physical, causal, mechanical nature that becomes thematic in the naturalistic attitude. In fact, Husserl never intends to place spirit into nature as one would do in the naturalistic attitude, but rather, he looks into the dimension of passivity as the *basis of spiritual activities* precisely *from the standpoint of spirit*. It is true that he makes use of the word "natural mechanism [*Naturmechanismus*]" (Hua IV, 338) in order to describe this dimension. But this word is used in quotation marks to express the passivity at the basis of spirit that originates or generates itself without any active participation of the Ego. Thus, this notion is never to be understood in the natural-scientific sense. Moreover, later in section 61 of the published version of *Ideen II* in which the same theme is discussed,[7] Husserl definitely states that the "soul" as the "underlying basis" of spirit is "not an objective (natural) reality" defined "psychophysically" as "a real unity in relation to circumstances of objective nature." The rules "to the formation of dispositions as substratum for the position-taking subject," for example, those "of associative dispositions," cannot be directly treated as "psychophysical." The soul and its "immanent lawfulness" *never* belongs to "the world of the mechanical, the world of lifeless conformity to laws [*die Welt des Mechanischen, der toten Gesetzmäßigkeit*]" (Hua IV, 279–280).

260

Nature as the underlying basis of the spirit that Husserl is describing here is, in my opinion, rather to be understood as a passively functioning constitution as the basis of the spiritual Ego: This basis functions passively without any participation of the Ego's rational activities, but continually forms a "history" of itself. The basis is therefore a dimension of *passivity* with which the Ego as the active spirit is never concerned. It is at the same time a dimension of *sensibility* that precedes and supports all rational activities of the Ego. It may also be considered as a *living nature* forming itself naturally, and continually sedimenting all its life as history. Husserl also calls it "my nature" (cf. Hua IV, 280), which "develops itself" and "organizes itself in its development" (Hua IV, 339). It is precisely this nature that gives the world as the ground of all egoical activities, namely the surrounding world [*Umwelt*], to the Ego.

As has already been mentioned, however, Husserl divides the natural basis of the spirit into the stratum of authentic, spiritless and that of unauthentic, spiritual sensibility. If this is the case, then we can say that, as a ground stratum of or under the surrounding world, Husserl also had in mind what could be called "primal nature [*Urnatur*]," given through the stratum of authentic sensibility. This giving stratum of sensibility would be, as it were, a natural basis of the spirit at the very beginning of its self-development and self-organization, and the given "primal nature" would be a ground stratum of the surrounding world before any sedimentation of spiritual products. This consideration leads us to the conclusion that Husserl here focused also on the natural basis of spirit before any sedimentation of spiritual products and correlatively on the primal nature given through it.

But it should be noted that Husserl in this text goes on to say that "the natural basis of the soul organizes itself in its development in such a way that 'nature' is constituted in it" (Hua IV, 339). This "nature" (in quotation marks) is, in my opinion, no longer primal nature, nor the surrounding world given in the self-development of that natural basis. Thus, the relation of the self-developing natural basis of spirit to the "nature" constituted in it now becomes problematic. I will attempt to give an account of Husserl's further analysis.

The way in which the natural basis of the soul organizes itself in its development and how "nature" is constituted in it runs as follows: At first, "the Ego behaves in general in its reactions as mere nature, and thus an 'animal' and purely animal Ego develops." But then, "for the [pure] Ego as active subject of the *cogitationes*, which is identically one throughout all of them, is constituted a new pregivenness: the empirical Ego." Since this Ego now "has a familiar nature, i.e., a nature to be learned in experience [= that 'nature' (in quotation marks)]," it is clear that the empirical Ego "has been generated in its natural evolution [= in the self-development of the natural basis of the Ego], precisely with its nature [= with that 'nature' (in quotation marks)], and that understandably purely according to 'natural laws' [= for example, the laws of association and reproduction pertaining to the natural basis of the Ego]" (Hua IV, 339). Thus, in the self-development of the natural basis of spirit, namely, in the self-organization of the natural basis of the soul, the empirical Ego with its "nature" (in quotation marks) is constituted for the pure Ego.

However, what precisely is the meaning of "nature" in this context? Husserl goes on: "in the constituted nature, the Body [*Leib*] and the Body-soul [*Leibseele*] are constituted as unity and [. . .] the empirical Ego is the Ego of Bodily-psychic nature [*Ich der leibseelischen Natur*]." This is "the Ego of the soul," which is "the constant subject" in its "mere re-acts, natural reactions in face of the possessions" (Hua IV, 339).[8]

Taking into consideration the employment of such words, "Body-soul," "mere re-acts," and "natural reactions," which Husserl uses to describe the naturalistic attitude, there is no doubt that he takes "nature" (in quotation marks) here as the nature to be regarded in the naturalistic attitude, i.e., as physical, causal, mechanical nature. In section 61 of the published text of *Ideen II*, he even asserts that the "natural side" of spirit, the "underlying basis of subjectivity," i.e., the living nature as the natural basis of the spiritual Ego, is indeed the "place of the constitution of a world of appearances, or of appearing objects," but that the constituted world is "the world of the *mechanical*, the world of *lifeless* conformity to laws" (Hua IV, 279; my italics).

Based upon the previous considerations, the following three observations can be pointed out: 1) In this text from the end of January 1917 Husserl focused on a notion of living nature as the basis of spirit, which is totally different from the physical, causal, lifeless nature to be grasped in the naturalistic attitude. The natural basis of spirit is the dimension of passive constitution at the bottom of the Ego as spirit. It functions without any active participation of the Ego, and develops and organizes itself, continually forming its own history. It is precisely through this dimension that nature as the surrounding world is given to the Ego. This Ego as spirit relates to [*sich verhalten zu*] this world. Without its natural basis, it could not perform its free, rational activities.[9] But recognizing the stratum of authentic sensibility within the natural basis (namely, the stratum that the natural basis should have been before any rational activity can begin to function) Husserl correlatively looked at a *ground stratum* under the surrounding world, too. In other words, he looked at the primal nature, which should have been given before any sedimentation of spiritual products. We must keep this in mind.[10]

2) Husserl thought at the same time that the natural living basis of spirit appears as physical nature in its self-development, in other words, that it is constituted in its self-organization as lifeless mechanical nature. What is constituted is neither a primal nature, nor a surrounding world, but a nature in the sense of the natural sciences.

3) But this does *not* mean that the natural living basis *itself* has to become lifeless nature in the natural-scientific sense. Husserl says, to be sure, that the empirical Ego with physical causal nature has been generated "in its natural evolution," i.e., within the self-development of the natural basis of the spirit (Hua IV, 339). But this empirical Ego is constituted as such precisely "for the [pure] Ego as active subject of the *cogitationes*, which is identically one throughout all of them" (Hua IV, 339). It is *for the pure Ego* that the natural living basis in its self-development comes to appear as physical, causal, lifeless nature. Precisely for the "spiritual regard [*geistiger Blick*]" (cf. Hua IV, 334) of the pure Ego, the natural basis is constituted as a psychophysical unity of the Body-soul.

We will see later in Section V that one can speak of spirit's "forgetfulness" of nature regarding this point. In the next section, however, I will first inquire into the process of the development of the idea "natural basis of the spirit which develops itself" in some other unpublished manuscripts of Husserl.

IV. The origin of the concept "natural, living, self-developing basis of spirit"

Monadological considerations of 1908–1910

At this point I will deal with some parts of the posthumous manuscript B II 2 entitled "Absolute consciousness. Metaphysical [consciousness]." (B II 2, 1a), the first 11 papers of "1908 or 1909" entitled "Monadology" (B II 2, 3a–13b), and the following 4 papers from "1910" written as a supplement to the former (B II 2, 14a–17b). On the basis of the belief that "consciousness cannot begin and consciousness cannot end" (B II 2, 12a), Husserl here considers, according to monadology, a "possibility of the origin of the souls and the meaning of the nature before 'creation of the souls' " (B II 2, 12a). I will summarize his descriptions in brief.

According to the "monadological view," "the monads are [or exist] from eternity, firstly in a 'state of slumber,' and then awake gradually" in such a way that "organic nature" is created (B II 2, 12b). Thus, "nature before all organisms, 'nature before the occurrence of consciousness' means [. . .] that all monads were in a state of slumber, that of 'involution' " (B II 2, 14b), and "the first development of organic entities means [. . .] the first awakening of monads" (B II 2, 14b). For human spirits or consciousnesses as "awakened monads," the "In-itself of the nature [*An-sich der Natur*]" means only "a rule concerning the course of possible sensations and apprehensions" (B II 2, 15a). But according to the monadological view, this rule is to be taken as "a rule for all monadic changes" (B II 2, 16a), as "a rule for all monads" (B II 2, 16a) including "all the sleeping monads" (cf. B II 2, 16b). Accordingly, "nature before all awakened consciousness means [. . .] that a regularity exists which develops the monads [= all sleeping monads] to 'waking' consciousness" (B II 2, 16b–17a). After all, "there is nothing other than all the monads" (B II 2, 16b). Husserl maintains "that there is nothing other than 'spirits' in the widest sense" (B II 2, 17a).

On the basis of these statements, it is clear that according to the monadological view, Husserl here conceives of nature before all waking consciousness, the In-itself of nature, as monads in the state of slumber, as a *living nature* harboring in itself a possibility of self-development towards waking consciousness. Certainly, it should be noted that Husserl's main viewpoint here is, as the title of the manuscript suggests, a "metaphysical" one, performed *beyond* his phenomenological method which was first established in 1905–1907. If Husserl would have followed his method faithfully, he would have only been able to elucidate how the In-itself of nature [*An-sich der Natur*] appears to our waking consciousness, and what kind of accomplishment [*Leistung*] of consciousness functions behind this appearance. The "phenomenological analysis" could have only clarified that "the In-itself of nature means for the awakened monads [. . .] a rule concerning the course of possible sensations and apprehensions" (cf. B II 2, 15a), i.e., that nature as it is appears for human consciousness as nothing but that rule. Following the "monadological view," however, Husserl went beyond such a phenomenological analysis here, and he reached the idea of a *living, self-developing nature* comprising all the monads.

It is now clear that those analyses concerning the self-development of the natural living basis of spirit that I went into in the last section can clearly be supported by these metaphysical, monadological considerations cited above. Indeed, the In-itself of nature that appeared as "a rule concerning the course of [. . .] sensations and apprehensions" was, in these monadological consideration of 1908 and 1910, not yet described as a dimension of passivity belonging to the basis of spirit. But it was, although put in metaphysical terminology, already discerned that the In-itself of nature is a *living nature* harboring in itself a possibility of *self-development* toward a waking consciousness. I think that the monadological consideration of 1908–1910 in B II 2 must be seen as a background of the phenomenological analysis dated at the end of January 1917 in the supplement XII of *Ideen II*, the analysis concerning the natural, living, self-developing basis of spirit.[11]

As was already mentioned in the second section, however, Husserl maintained in the original manuscripts of *Ideen II* (1912–1913) a one-sided superiority of spirit over nature on the basis of the dualistic confrontation between nature and spirit. There he did not at all recognize nature as the basis of spirit. How, then, did he come to focus his attention on the natural basis of the spirit as the dimension of passivity in the later manuscripts of 1917? I will briefly try to reconstruct the course of Husserl's thinking in the development between 1912/1913 and 1917.

The way to the recognition of nature as the basis of spirit

As far as one can tell from Husserl's manuscripts dated 1912–1913 (A VI 10, 3–5), which are related to the original manuscripts of *Ideen II*, his main aim was to distinguish the personalistic from the naturalistic (natural-scientific) attitude, and he only emphasized the difference between spirit (or person) and nature. There, for example, he states that spirit alone is a true, unrepeatable individuality while physical and mathematical nature is repeatable as many times as possible (cf. A VI 10, 4b–5a). Noticing the difference between both attitudes, he indeed distinguished "nature as surroundings [*Natur als Umgebung*]" (A VI 10, 3b), that is, "the nature directly given in sensuous appearances and appearing with sensuous qualities" (A VI 10, 4a), from the "nature of mathematical physics" (A VI 10, 4a). But he only laid down this difference and did not yet speak of a notion of *nature at the bottom of spirit*. He only described spirit from the viewpoint of its opposition to nature, and went on to analyze the superiority of spirit over nature and that of the personalistic attitude over the naturalistic.

To the best of my knowledge, the beginning of his attention to the natural basis of spirit is to be found in a text written approximately in December 1914 or a little later,[12] a text concerning the passive Ego (cf. A VI 10, 19–21). Husserl says here: "Opposed to the active Ego stands the passive, and the Ego is always passive or receptive at the same time whenever it is active" (A VI 10, 20a: cf. also Hua IV, 213). Whenever the Ego performs an intentional act, it already "experiences stimulations" and eventually "follows" their "attractive pull [*Zug*]" (cf. A VI 10, 20a). This description clearly shows that Husserl has already reached the idea that a dimension of passivity or receptivity always already functions at the bottom of all egoical activities. And now, also taking into account that this functioning goes on *naturally* before all authentic activities of the Ego, the passive dimension can be called "nature," although in a completely different sense from the objective nature on one hand, and from the nature as surroundings of the spirit on the other. I suppose that this was a motive which led Husserl to describe the passive dimension at the bottom of spirit as *natural* side or *natural* basis in the text of January 1917. And I think that his insight into this natural, passive dimension at the bottom of spirit was one of those motives that lead Husserl from his static to genetic phenomenology.[13]

Up to now my account has roughly shown the origin and background of the idea of "a natural, living, self-developing basis of spirit." In the following section, I will go into the problem that I hinted at in the third section, the problem of "spirit's forgetfulness of nature."

V. Spirit's double forgetfulness of nature

As I explicated the idea of "natural basis of spirit" from Husserl's text in the third section, we found at the same time that he describes in his text that the natural basis in its self-development becomes constituted as a physical, lifeless, mechanical nature, and yet precisely for the spiritual regard of the pure Ego. It seems to me that Husserl has elucidated there a kind of "spirit's forgetfulness," that of its natural living basis, probably even without realizing it.

Upon reconsideration, however, it could already be seen that Husserl clearly speaks of spirit's forgetfulness of its nature in another text as well. For it had been mentioned in the second section that Husserl in the original manuscripts of *Ideen II* describes that the naturalistic attitude is based on a kind of self-forgetfulness of the personal Ego, namely, on the spirit's forgetfulness of its nature as the surrounding world.

It seems, then, at first sight that the previous investigations have shown the following two modes of spirit's forgetfulness of nature: (1) spirit's forgetfulness of its natural, living, self-developing basis, (2) spirit's forgetfulness of its surrounding world done by the spirit in the naturalistic attitude.

But a more detailed consideration will now be necessary. For, taking into account that nature for spirit, its surrounding world, is given through its natural living basis, one could take those two kinds of forgetfulness sketched above as aspects of *one forgetfulness*, namely the situation that the human spirit, by taking the naturalistic attitude, comprehends the whole world including its body and soul as physical, mechanical nature, so that it forgets its natural living basis and at the same time its surrounding world given through it as well. In fact, Husserl's critique of Modern natural science and his call for the return to the pre-scientific lifeworld [*Lebenswelt*] in *Krisis* could be interpreted as a severe critique of such forgetfulness. If one could interpret those two kinds of forgetfulness in this way as suggested, Husserl's earlier analyses could be seen as the seed of the lifeworld problematic in *Krisis*, and,

[41]

more generally, as a phenomenological elucidation concerning the Modern human spirit being accustomed to a natural-scientific point of view.

However, if we also take into consideration that Husserl, by recognizing a stratum of authentic sensibility within the natural basis of the spirit, had in mind a primal nature as a ground stratum of the surrounding world, then we must by all means distinguish spirit's forgetfulness of its own natural basis from that of its surrounding world. For, by forgetting its natural basis, spirit forgets not only its surrounding world – the world given through the basis in its self-development –, but *must have also already "forgotten" primal nature*, which should have been given at the very beginning of that development. Besides, it is no longer possible to relate spirit's forgetfulness of nature directly to the naturalistic attitude, for the "forgetfulness" of primal nature, which should have already begun with the sedimentation of authentically spiritual activities, does *not necessarily* belong only to the Modern human spirit accustomed to this attitude. Accordingly, we must distinguish two sorts of forgetfulness. But it is not correct to call them spirit's forgetfulness of its natural basis and that of its surrounding world, as one would expect. They should rather be called spirit's forgetfulness of primal nature and that of its surrounding world. I shall explain this point in detail.

Husserl's analyses in his last years have clarified that the surrounding world or the lifeworld – the nature to be given in the personalistic attitude (as the authentic natural attitude for the spirit) – is a nature which is *always and already passively typified* before any authentic activity of spirit.[14] Since such typification of the lifeworld must have originally been dependent only on the living body [*Leiblichkeit*], and therefore on the laws of sensibility, the typified lifeworld must have been a plain, authentic nature (= primal nature) for spirit, as long as sedimentations of spiritual activities did not yet reach into the dimension of sensibility. From the beginning of these activities, however, the *logos* of spirit, reason, always already participates deeply in the process of passive typification of lifeworld, and thus history and culture as passive sedimentations of spiritual activities are inseparably combined with this process. Therefore, the surrounding world or lifeworld as typified nature can no longer be primal nature in the sense of a spiritless one.[15] And in principle, this had also been the case before nature in the natural-scientific sense was constituted by the Modern spirits. But now we can say on the other hand that the typified lifeworld with sedimentations of rational activities of spirit is what it *must* be for self-developing spirit, and that it is in this sense nothing but an "authentic" nature for the spirit. For the self-developing spirit cannot return to the primal nature any more, as long as it remains in its natural, personalistic attitude. It is precisely in this sense that it is possible to speak of a "forgetfulness" of primal nature that the spirit as such must do, distinguishing it from that of the surrounding lifeworld which the Modern spirit does in its natural-scientific point of view: In the self-development of its natural basis, and yet through the sedimentation of products of its own activities into this basis, every spirit inevitably "forgets" the nature which this basis should have given before all spiritual sedimentations (= the spirit's forgetfulness of primal nature). And the Modern spirit accustomed to taking the natural-scientific viewpoint, furthermore, views the whole world including its body and soul as a physical, mechanical nature, so that it also forgets its own natural living basis and the surrounding nature (lifeworld) which is given to it through the basis in its self-development (= Modern spirit's double forgetfulness of primal nature and of the surrounding lifeworld including its own living body).

Still we are left with one question: Why does spirit "forget" nature? And why does Modern spirit forget nature doubly? It is not easy to understand thoroughly the motives of spirit. I shall suggest a motive only roughly, even though it will probably exceed the boundaries of Husserl's texts.

Spirit inevitably forgets primal nature, and Modern spirit in the naturalistic attitude further forgets its living body and its surrounding world in their "natural" givenness as well. For Modern spirit, accordingly, nature is hidden or veiled in the double sense. But essentially, the primal living nature (*physis*) cannot hide itself. It must always be as it is or generates itself. Therefore, it is not the case that nature is "accustomed to hide itself,"[16] but rather that *spirit hides it in the way of forgetting it.*

But why does spirit "hide" nature in the way of forgetting it? One motive for this, I suppose, is that spirit, especially Modern spirit, has a strong tendency to posit itself as an *intellectus agens*, as subject of its purely reasonable activities, and to act as *pure reason*, not only in its scientific work but also in its daily life. In fact, Modern natural science was nothing but a rational interpretation of the entire world by the spirit that was conscious of itself as pure reason. Here, for example, Descartes's discovery of *ego cogitans* through his methodical doubt and his recovery of the extensive world should be taken into account. As we have already seen, however, spirit has, as its basis, a natural side to be called "passivity," "instincts," or "feelings," the existence of which it must recognize. Will not spirit as pure reason veil this natural side in the way of forgetting, precisely because this side can never be scooped up with the *logos* of the spirit, namely with reason? It seems to me that there is a strong case for suggesting that various repressions and violence done to nature, animals, children, the insane, and the non-european especially since the Modern age[17] have been brought about by spirit positing itself as pure reason.

VI. The significance of phenomenology for our time

What can phenomenology do in the face of spirit's forgetfulness of nature as reconstructed here? Can phenomenology be expected to do anything at all, even though phenomenological description itself is one of logical and rational practices carried out by the *logos* or reason of the human spirit? In conclusion, I would like to make some comments on this topic.

The previous considerations have first shown that spirit, according to Husserl's *phenomenological analyses* concerning the constitution of nature and spirit, is neither completely separated from nature, nor one-sidedly superior to it. We have rather seen that a passive dimension of natural living basis of the spirit is taken into consideration in his texts. Then, we have discussed that these phenomenological descriptions elucidate, consciously or unconsciously, spirit's double forgetfulness of nature, namely, that of primal nature which must inevitably be done by spirit as such, and that of the surrounding lifeworld done by Modern spirit accustomed to the natural-scientific perspective. Accordingly, it can and must be said that it was precisely through Husserl's phenomenological analyses that spirit's double forgetfulness of nature could become apparent.

Generally speaking, phenomenology may be considered as a scientific endeavor which tries to reach insights into implicit, anonymous functions of consciousness by looking at the differences between the really immanent givenness and the outer appearance for spirit living in the mode of the natural attitude. And in order to return to the really immanent data from the outer appearance, it will suspend all the judgment and positings of the natural attitude. In fact, in a number of phenomenological analyses in and after the time of *Ideen I*, in which Husserl occupied himself with the constitution of nature and spirit, Husserl elucidated various implicit, passive functions and habitualities of consciousness as the natural basis of spirit and also the forgetfulness of nature carried by spirit. It seems to me that his analyses thereupon became one of those motives which lead Husserl from static to genetic phenomenology. His later genetic phenomenology was indeed a persistent effort to approach and describe the passive dimensions of consciousness that are found in this paper to be the natural living basis of spirit. It was a persistent trial to reach insights into the fact that there certainly is a natural basic dimension that can never be adequately scooped up with the *logos* of the spirit.

To be sure, it is impossible for phenomenology, as one of logical and rational practices of human spirit, to transcend the *logos* of spirit or to free itself from history, culture, and tradition as sedimentations of spiritual activities.[18] Thus, phenomenology is a constant effort to logify what is never to be logified adequately. It must face various bounds in the course of its descriptions. Nevertheless, phenomenology is precisely a *philosophical* practice of human spirit, if philosophy means that human spirit turns its reflective eye back again to the matters themselves that remain implicit or anonymous in its everyday life. If this is the case, and I believe it is so, it is necessary for present humanity to go ahead with this philosophical practice of phenomenology, at first following Husserl, and then going beyond him.[19]

Notes

1. Cf. Edmund Husserl, *Ideen zu einer reinen Phänomenologie und phänomenologischen Philosophie. Zweites Buch. Phänomenologische Untersuchungen zur Konstitution*, ed., Marly Biemel, Martinus Nijhoff, The Hague, 1952 (Husserliana Vol. IV), pp. XIII–XX. Husserl's works that are published in Husserliana are hereafter cited as "Hua" followed by volume and page number. In the quotations from Hua IV, I have generally used the following translation: Edmund Husserl, *Ideas Pertaining to a Pure Phenomenology and to a Phenomenological Philosophy, Second Book, Studies in the Phenomenology of Constitution*, translated by R. Rojcewicz and A. Schuwer, Kluwer Academic Publishers, Dordrecht/Boston/London, 1989.

2. Husserl's unpublished manuscripts used in this paper are cited according to archival organization and documentation found in the Husserl-Archives in Leuven.

3. The "original pencil manuscript of *Ideen II*" ("Ursprüngliches Bleistiftmanuskript von *Ideen II*," written in October to December 1912) and the "central manuscript of the third section of *Ideen II* (The constitution of the spiritual world)" ("Hauptmanuskript zum dritten Abschnitt der *Ideen II* (Konstitution der geistigen Welt)": the so-called "H-Blätter" of 1913). The former is now preserved in the Husserl-Archives of Leuven under the signature F III 1, 5–36. The latter is to be found there in the convolute M III 1 I 4. In the quotations from the "Blätter," I will, as an exception, give first the original page number of the papers (with the signature "H") and then in brackets the present signature and page number of the Husserl-Archives. Concerning the original manuscripts of *Ideen II* cf. also Tetsuya Sakakibara, "Das Problem des Ich und der Ursprung der genetischen Phänomenologie bei Husserl," *Husserl Studies*, 14(1), 1997, pp. 21–39.

4. According to M. Biemel, the text of the section 64 of the present *Ideen II* was formed in its final version approximately in the years of 1924 and 1925, on the basis of E. Stein's second elaboration of 1918 (Cf. Hua IV, XVIIIff.).

5. Cf. Hua IV, 334. In the central text of *Ideen II*, Husserl also calls these sedimentations "secondary passivity" (cf. Hua IV, 12, 20). The secondary sensibility or passivity is made possible only by retentional consciousness. In a supplement of *Formale und transzendentale Logik*, Husserl speaks clearly of a retentional sensibility as the first form of secondary sensibility (cf. Hua XVII, 319ff.).

6. The "laws of association and reproduction" are "more general ones, which extend beyond primal sensibility" to the stratum of unauthentic, spiritual sensibility (Cf. Hua IV, 337).

7. According to M. Biemel, the text of section 61 was formed approximately in the years of 1924 and 1925 on the basis of E. Stein's second elaboration of 1918 (cf. Hua IV, XVIIIff.).

8. In the development from the Ego of the soul to "the free personal subject," not only "laws of association," but rather especially those "of reason" must determine the Ego (cf. Hua IV, 339–340).

9. Husserl writes: "[. . .] each free act [of the Ego] has its comet's tail of nature" (Hua IV, 338).

10. Two years later in his lecture on "Nature and Spirit" from February 21, 1919 and his manuscripts regarding this lecture [*Aufsätze und Vorträge (1911–1921)*, eds., Thomas Nenon
and Hans Rainer Sepp, Hua XXV (Boston: Kluwer, 1987), 316–330], Husserl explicitly
speaks of "nature" that is "sensuously given" before all spiritual Ego-acts (329): In order
to ground the distinction between natural and human sciences, Husserl analyzes there
the "stratification [*Schichtung*] of the pre-given surrounding world" (325), and through
a "reductive" method of "dismissing [*abtun*]" (326) all "meaning-predicates" belonging
to the "surrounding objects" (325), Husserl discovers the stratum of "natural Objects in
themselves [*Naturobjekte an sich*]" as "the lowest level" of the surrounding world (328).
This stratum is a "nature" that is "originally and perceptually given in pure receptivity"
(329) before any egoic act that gives spiritual meanings to those objects. It is now clear
that this nature is 'primal nature' as a ground stratum of the surrounding world.

However, it should be noted at the same time that Husserl, in the course of this lecture,
also speaks of "nature" in a "new sense" (317). He says: Those "directly intuitive natural
Objects," to be sure, do not have any meaning-predicates arising from egoic acts, but they
do still have "sensuous qualities" relating to the "living body" of the respective subjects
(316, 329), so that these qualities must always be "subjective-relative" (316). Now it
was precisely "Galilean natural science" that brought "what remains invariant" out of
all subjective-relativities of the intuitive natural Objects and recognized it to be the "In-
itself of natural Objects in a new sense ['*An-sich' der Naturobjekte in einem neuen Sinn*]"
(317). This new 'nature in itself' "can no longer contain anything sensuous" (317). It is
rather a "mathematically exact nature" determined "only in pure logical and mathematical
predicates" (317). Against this nature, the "sensuously intuitive nature" is no more than
a "mere appearance" of that one (317). Thus, the "new natural science" substituted the
"mathematically exact nature" for the directly intuitive one as a ground stratum of the
surrounding world (cf. 317).

In his lecture-course on nature and spirit from the summer-semester 1919, Husserl describes "the ideal lowest level [*die ideell unterste Stufe*]" of the surrounding world as
"mere nature" and correlatively describes the "subject of mere nature" as a "subject of
possible, passively sensitive apperceptions, namely, those which gain their sense-contents
without any co-participation of spiritual producing" (F 135, 117a–b). In the lecture-course
on ethics from the summer-semester 1920 he also tries, through a method of "dismantling
[*Abbau*]," to go back from the "spiritual stratums of our surrounding world" to the "world
of pure experience" as a ground stratum (cf. for example A IV 22, 42b, 47bff., 54a). But it
should be noted here that there can be seen a tendency in these two lecture-courses to take
the ground stratum under the surrounding world as an abstract object of Modern physics:
Husserl himself has a tendency to regard this stratum as a physical, lifeless, mechanical
nature, and his spirit itself has a tendency to forget nature, the tendency which we will see
later in Section V. In my opinion, however, the problematic of the ground stratum under
the surrounding world and of the giving stratum of authentic sensibility leads directly to
that of "instinct" or "impulse-intentionality [*Triebintentionalität*]" in the latest Husserl.
In Merleau-Ponty, the natural basis of spirit, especially its ground stratum of authentic
sensibility, might be also called "*un esprit sauvage*" or "*l'esprit brut*," and "primal nature"
called "*un monde sauvage*" (cf. "Le philosophe et son ombre," in *Signes*, Gallimard, 1960,
p. 228). In this paper, however, I cannot go into this problematic in detail.

On those lecture-courses from 1919 and from 1920 cf. Guy van Kerckhoven, "Zur Genese
des Begriffs 'Lebenswelt' bei Edmund Husserl," in *Archiv für Begriffsgeschichte*, Vol. 29,
1985, pp. 182–203; Ullrich Melle, "Nature and Spirit," in Thomas Nenon and Lester Em-

bree (eds.), *Issues in Husserl's Ideas II*, Kluwer Academic Publishers, Dordrecht/Boston/London, 1996 (*Contributions to Phenomenology*, 24), pp. 15–35.

11. The fact that the monadological viewpoint belonged to the background of Husserl's phenomenological descriptions after 1908–1910 can be seen in his following descriptions: "[. . .] to consciousness itself belongs the unconditional essential possibility that it can become an alert consciousness, [. . .]. Or, to speak like Leibniz, that the monad passes from the stage of involution into the one of evolution and becomes [in higher acts] self-consciousness 'spirit.' " (F III 1, 5a (1912); cf. Hua IV, 108); "[. . .] This subject [= person] is conscious of itself and thereby is a developed subject at the stage of 'spirit' in Leibniz' sense." (Hua IV, 351 (1916–1917)); "To the pure essence of the soul there belongs an Ego-polarizing; further, there belongs to it the necessity of a development in which the Ego develops into a person and as a person. The essence of this development includes the sense that the Ego as person is constituted in the soul by means of self-experience." (Hua IV, 350 (Husserl's insertion into Landgrebe's version, after 1924/25)).

12. The page 19b of A VI 10 includes a writing dated December 7, 1914.

13. I cannot go into this point in detail here. Cf. Hua XIII, 346–357 (Beilage XLV, 1916/17); Hua XI, 336–345, "Statische und genetische phänomenologische Methode" (1921); Hua XIV, 34–42 (Beilage I, Juni 1921). The latter two articles will be included in Edmund Husserl, *Analyses Concerning Passive and Active Synthesis: Lectures on Transcendental Logic*, trans., Anthony J. Steinbock, Husserliana Collected Works (Boston: Kluwer Academic Publishers, forthcoming). They are available separately in English as "Static and Genetic Phenomenological Methods," and "The Phenomenology of Monadic Individuality and the Phenomenology of the General Possibilities and Compossibilities of Lived-Experiences. Static and Genetic Phenomenology," trans. Anthony J. Steinbock, "Static and Genetic Phenomenology: Introduction to Two Essays," *Continental Philosophy Review* (formerly *Man and World*), Vol. 31, No. 2 (1998). Concerning a first seed to the idea of genetic phenomenology in Husserl, see Tetsuya Sakakibara, op. cit.

14. Cf. *Erfahrung und Urteil*, Felix Meiner, Hamburg, 1985 (ph B 280), §8, pp. 26–36; §§81–85, pp. 385–408.

15. It seems to me that the ambiguities in Husserl's concept of the lifeworld that Claesges pointed out correspond to those in the idea of typified nature that I clarified here: The distinction which he made between the lifeworld in the narrower sense and the one in the widest sense might be compared to the difference between primal nature and the surrounding lifeworld with spiritual sedimentations. Cf. Ulrich Claesges, "Zweideutigkeiten in Husserls Lebenswelt-Begriff," in Ulrich Claesges und Klaus Held (eds.), *Perspektiven transzendental-phänomenologischer Forschung*, Martinus Nijhoff, The Hague, 1972 (*Phaenomenologica*, 49), pp. 85–101.

16. Cf. Heraclitus, Fragment 123.

17. Cf. for example Michel Foucault, *Histoire de la folie à l'âge classique*, Gallimard, 1972. Husserl's ideas of "europeanization of humanity" and of "teleology of european reason" in *Krisis* should also be remembered here.

18. Cf. Tetsuya Sakakibara, "Husserl on Phenomenological Description" (in English), in *The Ritsumeikan Tetsugaku (The Proceeding of the Philosophical Society of Ritsumeikan University)*, No. 6, 1994, pp. 1–18.

19. This paper is written as a result of my research in the years 1995–1996 in Wuppertal, Germany. I would like to thank the Alexander von Humboldt Foundation for the support of my research. I am deeply grateful to Professor Dr. Klaus Held, PD Dr. Dieter Lohmar, Professor Dr. Guy van Kerckhoven, Professor Shigeto Nuki, and Dr. Sebastian Luft for useful advice and comments on the earlier version of this paper. To PD Dr. Lohmar

272

and Mr. Michael Weiler who kindly helped me at the Husserl Archives in Köln and in Leuven, and to Professor Dr. Rudolf Bernet, the director of the Husserl-Archives in Leuven, who allowed me to cite Husserl's unpublished manuscripts, I owe a special debt of gratitude. Last but not least, I thank Mr. Mark Sainsbury for his kind help regarding English formulation.

Note about the author

Tetsuya Sakakibara (born 1958; B.A. and M.A., University of Tokyo, 1983, 1986) is Associate Professor of Philosophy at Ritsumeikan University in Kyoto. He has put much effort into interpreting Husserl's phenomenological method and researching his unpublished manuscripts. He is author of "Husserl on Time-Analysis and Phenomenological Method," *Japanese and Western Phenomenology*, 1993 and "Das Problem des Ich und der Ursprung der genetischen Phänomenologie bei Husserl," *Husserl Studies* 14(1), 1997. He is currently interested in the process of Husserl's thought as it progresses from static to genetic phenomenology, and the various problems of phenomenological description.

Continental Philosophy Review **31**: 273–291, 1998.
© 1998 *Kluwer Academic Publishers.*

The theory of association after Husserl: "Form/content" dualism and the phenomenological way out

SHIGETO NUKI
Department of Philosophy, Saitama University, Japan

Introduction

This paper will address one of the "unsolved" problems that has traditionally been called the "form/content" dualism.

Husserl writes in *Ideas I*: "This remarkable duality and unity of *sensory hyle* and *intentional morphe* plays a dominant role in the whole phenomenological sphere (in the whole sphere, namely within the stage of constituting temporality, which can be constantly verified)" (Hua III, 192). To be sure, Husserl has defined the sensory hyle "descriptively" as the really immanent [*reell*] part of consciousness with differences of color, sound, touch and so on, as well as "functionally" as something that can become an adumbration or an aspect of objects through the animating apprehension of active consciousness. But the constitution of an object in his later analysis is clarified more and more in terms of noematic unification (Holenstein, 93), which obscures the function of *hyle* in the process of constitution.

As is indicated in the parenthesis of the citation, the overwhelming dominance of the schema was explicitly rejected in the analysis of time consciousness. In a famous footnote of *The Lectures on Internal Time Consciousness* Husserl writes, ". . . not every single constitution has the schema apprehension-content of apprehension" (Hua X, 7). This statement gives rise to more irritation than solution. Because the formal constitution of temporality consists in the transition of an originary impression into the retention of it, it would seem that inside the temporal constitution is found, if not the dualism of "apprehension/content," then that of "temporal form/temporal content." This leads us to a problem similar to the one I have indicated above. At the end of his *Formal and Transcendental Logic*, Husserl criticized the "overall dominating datum-sensualism" that "builds up the life of consciousness from data, so to speak, as ready-made objects" (XVII, §107b, 251). It is quite difficult to see the originality of Husserl, however, if his only claim is that sensory data are constituted in internal time consciousness, as many interpreters suppose

(Holenstein, 198). In his *Analysen zur passiven Synthesis*, which treats the deepest stage of the constitution of sensory data, the sphere of the formal constitution of temporality is declared to belong to another dimension than that of passive, material syntheses (Hua XI, 125f). I think that these problems can be principally solved by appealing to Husserl's analyses concerning association in his *Analysen zur passiven Synthesis*. By introducing this phenomenon as a key concept, however, we are necessarily confronted with another difficulty that Derrida raised in *La voix et le phénoméne*.

Derrida has claimed that in keeping with Husserl's tendency toward the traditional metaphysics of voluntarism (VP, 37), Husserl tried to reduce everything to the sphere of expression or the active life of intentional consciousness by 'reducing' indicative signs; despite this attempt, claims Derrida, inside the very core of Husserl's phenomenology – in concepts such as temporality, reflection or intentionality – is in reality the hidden structure of *différance*, which turns out to be the residue of indicative signs.

The concept of association appeared first in 1901 as the psychic origin of *indicative signs* (LU, 1, §4). At least up to 1921 Husserl regarded this phenomenon not as transcendental, but as empirical (Yamaguchi, 88). We find his locution about association as "a kind of causality" in a text, which is supposed to be written in 1920 or 1928 (Hua XI, 386, 1. 5f). Although Husserl or his interpreters have claimed repeatedly that the phenomenological concept of association is something 'regulated' (Hua XI, 117), 'intentional', 'a priori' (Holenstein, 3) or even 'transcendental' (Yamaguchi, 14–16, 88), in contrast to its psychological counterpart as 'empirical', 'causal' or 'blind' mechanism (Held, 1986, 23), we cannot deny the fact that association is an indeterminate or accidental occurrence, in contrast to this necessity of signification.

Husserl's most comprehensible formulation of association reads: "Something present recalls something past" (Hua XI, 118). As this formulation indicates, this phenomenon of association has been in the first place analyzed as, or in, the mechanism of remembering. If a young person standing before you reminds you of his father, an association works between the face of the young man and that of his dead father. To put it phenomenologically, for intentional life, which is necessarily oriented toward the future to fulfill empty places in present perception, it is impossible to step backwards into its own past solely by being motivated through its intrinsic mechanism. In order to explain the emergence of backward intentionality, Husserl has adopted a totally different kind of principle from active cognition, namely association, which therefore can be regarded as the condition for the possibility of remembering.

The problem lies in the fact that even though association enables remembering, it does not guarantee the certainty of the past (Hua XI, 193, 1. 33–35). A memory can easily be erroneous because a present datum (P) can awaken more than one past datum, and these are similar to P in one respect or another. The possibility cannot be ruled out that more than two past data are blended or confused, thanks to the affective associations that occur between these data (ibid. 194–195). Consider the case where you confuse the memory of a journey you took last year in Idaho with the one you took four years ago in Ohio. To make matters worse, you cannot even ask the memory at hand if it is 'correct', because you have no access to your own past other than remembering. 'Objective' testimony cannot help either, without being interwoven into a remembering structured in a narrative form, for otherwise you could not know 'of what' it is a proof. A memory can be known to be false only by means of another, more distinctive memory of the same past, but this cannot be acquired until after the fact.

The unreliability of memory is due to the fact that association always occurs accidentally: First, a present datum always awakens a past according to a certain aspect of resemblance between the two. It is always indeterminate and accidental which aspect or part is chosen by an associative awakening (Hua XI, §26, 32, p. 427). Secondly, because there can exist more than two past data which are similar to a present datum in a certain respect, it is decided only by chance which past datum is awakened by the association released by a specific aspect of a present datum. You cannot even foster an association or forbid it to happen. Brand was quite correct in saying that association is similar to the folk tale figure of "*Siebenmeilenstiefel*" (Brand, 108, cf. Holenstein, 67 f). You yourself cannot choose where to go, or even whether to go somewhere with the boot: This is why association is regarded as a *passive* phenomenon.

How is it to be understood that such a problematic phenomenon is introduced in phenomenology as a "transcendental-phenomenologically fundamental concept" (Hua 1, 113–114)? In the following sections, I will examine the phenomenological analyses concerning association (I), in order to see if and how it has solved the form/content dualism problem (II). Through these observations the role of association within perception appears more and more enigmatic. After fixing the status, not of association itself, but of the considerations concerning it in the theory of intentionality as a whole (III), and evaluating the question of *hyle* (IV), I will turn to the criticism of Derrida, although in the context of this article my treatment of Derrida will have to be rather cursory (V).

I. Preliminary analysis concerning association

Despite Husserl's tendency to conceive of temporal constitution as isolated and primary in contrast to the rest of constitution, which even made him assume the immortality of transcendental ego (Hua XI, 380, Cf. Hua XV,

379–380, manuscript, C-4/11, Nuki, 1989, 162–163), we can admit that the constitution of temporal loci presupposes material synthesis: The temporal locus of a past event is fixed only in a chain of past events, which is constituted by following the associative indications of the remembered up to the present of remembering (Hua XI, 193). Retention as a "getting away from the present [*Entgegenwärtigung*]" (Fink, 23, Held, 1966, 28) cannot afford us temporal order. Husserl writes: "Every retentional process loses its affective force in the change, and finally loses its life" (Hua XI, 170). The internal fusion of formal and material constitution can be observed more clearly by examining the constitution of sensory unity in the living present.

Husserl's analyses of association inside the living present, including the retained temporal phases, can be divided into several stages. After picking out the rules of association obtaining between already constituted units through an observation that he calls a 'static observation' (Hua XI, 130, 1. 8) – rules of association such as resemblance (Hua XI, §28), temporal or spatial order (Hua XI, §29) – Husserl undertakes a 'kinetic observation' (ibid., 1. 13) in which he depicts the process of synthesis between more than two unified consciousnesses of sensory units: When, e.g., two data, A and B, are similar to each other, A affects the ego, and if the ego pays attention to A and B, affection is released, so that the coincidence or overlapping between the consciousness of A and that of B takes place, on one hand, and A and B appear as a pair, on the other (Hua XI, 130 f). Here, Husserl takes a point of view quite near to a naturalist who presupposes the resemblance of sensory units, though the ego at work cannot speak of it in reality before paying attention to it. Not only do sensory units appear as something already made, but also the formal condition of temporality remains outside the units or the process of synthesis.

It is in the third stage of analysis that the most important insights of Husserl show up (Hua XI, 137–177). In the first half of this stage, where the process of building prominent unity itself is put under the loupe of phenomenology, he stresses the inner fusion of temporal continuity and material unity again and again. He writes: "In other words, every prominent datum is not merely juxtaposed with other data in the living relations of succession. Rather, it has *in itself* an inner synthetic structure and in particular is *in itself* a continuity of sequence" (Hua XI, 140, italics mine). The founding of material unity through temporal succession can be seen more clearly in the following sentences of Husserl: "This inner continuity is the *foundation* of a continual fusion [*Verschmelzung*] with respect to content, fusion at-close-proximity" (*ibid.*, italics mine), or: "A certain concrete unity as the unity of immanent givens is only conceivable as the continuity of content *in and through extension* as the continuity of duration" (Hua XI, 141, italics mine).

The reason why a concrete, prominent unit is possible exclusively on the basis of temporal duration becomes clear when we examine the next step of analysis where Husserl introduces the notion of 'affection' (Hua XI, chap. 2, §32, p. 148 f) and 'awakening' (Hua XI, 152–153).

Husserl initially characterizes affection as an asymmetrical relationship of "the allure in the intentionality of consciousness [*bewußtseinsmäßige Reiz*], the peculiar pull that an object given to consciousness exercises on the ego" (Hua XI, 148), while 'awakening' can be regarded as a symmetrical relationship between more than two sensory units. The boundary between the two notions becomes quite obscure at times in the course of his analysis (Hua XI 154, 172), but the problem with which Husserl deals here, and which I think forms the central question of the whole analysis, concerns the phenomenological order of priority between affection, awakening, and the unification of sensory data. The question is important because in answering it his notion of association undergoes complete transformation.

Up to the second stage Husserl thought that affection or awakening takes place only *after* sensory units have already been formed, for association was conceived as a mechanism to build a relation as the descriptively higher form of unity since the *Logical Investigations* (LU, I, 29). In the case where we discover the resemblance between Louis the 15th and a pear, this relationship of resemblance is realized as a unitary form after the person and the fruit respectively have been recognized, where the mechanism of forming units as the related terms can be taken for granted. Even after the entry into the second half of the third stage he writes at first: "Affection presupposes prominence" (Hua XI, 149).

In the course of examining the mechanism of forming units Husserl has discovered the dynamic relationship of awakening as *conditio sine qua non* for the entire phenomenon. Husserl writes: "In the final analysis, does not the essential, lawful regularity of the immanent formation of unity that we have described . . . express the mere conditions of the possibility of such unities, *while the actual emergence of these unities itself is dependent upon affection and association?*" (Hua XI, 153, italics mine). This insight that Husserl indicates here arises from the observation of cases in which the 'transmission of awakening' from another unit is required for a unit to be formed. Husserl introduces several illustrating examples: Consider the case where you hear a tone sound that is initially pretty loud, and then becomes softer until you can no longer hear it (Hua XI, 152–153). The last *pianissimo* tone has no clear contour any more and would get lost in the entire acoustic field if the phase did not proceed it. It affects you only because the transmission of awakening radiating from the loud tone is passed over to it, which and which alone makes it possible for the last tone to be a sensory unit. Another example shows how strong this transmission of awakening can be: While occupied by something else you sometimes do not realize a melodious sound. Only after an especially moving air appears, it affects you so much that the past melody as a whole becomes prominent, in which case the transmission of awakening works backwards into the retained phases of your consciousness (Hua XI, 155).

These examples show us that we cannot speak of unification without the transmission of awakening. It is also impossible to treat unification and affection separately, because all that matters here is the *sensory* unit, but not unity in general. You can surely say of a water drop – a favorite model for the psychologists – that it is a pregnant Gestalt even if no one regards it, but it makes no sense to say of a *sensory* color or figure that it is a unity without anyone being aware of it. Husserl writes at the end: "What is constituted in the intentionality of consciousness exists for the ego only insofar as it affects me, the ego" (Hua XI, 162). We must thus conclude that *affection, awakening* and *unification* are three aspects of one and the same phenomenon, which we cannot consider in isolation, and I think that this conclusion leads us to dispense with the form/content dualism.

II. Form/content problem and its solution

It will suffice for this purpose to show that material synthesis is inconceivable without the formal constitution of temporality, on one hand, and that temporal form is impossible without taking into account material synthesis, on the other.

By the unification of a black spot on a white paper, we cannot talk about the difference between form and content, to be sure, because each sensory unit emerges thanks to contrast or the figure/ground relationship between them, but not *vice versa*. Welton was correct in saying that "while there is a distinction between synthesis and what is formed in the synthesis, the relative independence of 'content' and apprehension does not exist at this level [of associative synthesis]" (Welton, 63). Husserl would admit this, too. But, this explanation does not allow us to abandon the dualism of *temporal* form/content.

Reconsider the first example of acoustic perception. In order that a unit in a present phase (2) is formed, a (transmission of) awakening from the advanced phase (1) is required, so that a perception of a durable tone takes place. It makes no sense to speak of this kind of (transmission of) awakening if it lacks the temporal difference between the two units (1) and (2). True, we can talk of awakening within one and the same temporal phase, however it brings about not a durable tone but rather an instantaneous one. Even for

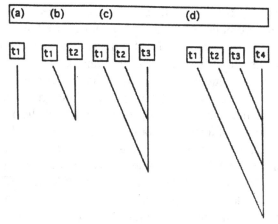

Figure 1.

an instantaneous tone to be recognized, an awakening in the form of contrast between the phases with and without tone is necessary. We cannot regard material synthesis as something separated from formal synthesis. This is the very reason why Husserl has written: "A concrete unity, the unity of an immanent datum is only conceivable as the continuity of content *in and through* the continuity of an extension as the continuity of a duration" (Hua XI, 141) *et cetera*.

Now, it might seem quite difficult to demonstrate the second half of our proposition that temporal form is impossible without material awakening. Affection or awakening cannot be regarded as something that *causes* or enables temporality. We can even conceive of the formal constitution of time in isolation, which we shall observe more closely.

Husserl's theory of time is laden with a kind of 'ambiguity': It admits the enlargement of intuitive presence into near past, which takes place in and against the process of 'getting away from the present'. As against the active intentionality of perception or remembering, which pursues the truth by integrating the evidences about a perceived object, retentional continuity has the character of diffusion. The famous diagram of time illuminates this quite well (see Fig. 1). Although original impression and original consciousness (a) which occur in T1 pass away as soon as a next 'now' (T2) appears, T1 does not vanish altogether, because the original consciousness of T1 is retained in the phase of T2. This modification is expressed in the diagram by diagonal line from T1 to T2 (b). As soon as T3 appears, the phase of T2 passes away, but the complete phase of T2 is retained in the phase of T3 (c), and so on (d).

Because the diagram itself is a product of the 'spatialization of time' and the triangle as a whole cannot exist in our time consciousness, it is only the phase of T3 which exists at the time when the proceeding T1 and T2 are already past. The past points (T1, T2 . . .) are not glanced over in one central, privileged perspective, but are retained from different *termini a quibus* which are lined up in the phase of T3. Remember another diagram of the earlier Husserl, where the retentional intentionality radiates out of one point of the present surveying all the temporal positions in a single glance as they remain at the points where they were originally given. Moreover, the object intentionality of the perception of a melody, for example, oriented to the coming tone in the future, does not have the retained past as its object. A past time-point or datum can be an object only in remembering. Husserl writes: "Time is known as object only in iteration (of remembering)" (Hua XXIX, 4). Thus, retention cannot but be regarded as the process of 'getting away from the present'. Even though the present field is temporally organized, this order itself cannot be objectified as such, but is only lived.

All these observations allow us to rethink the relationship between temporal form and material synthesis.

First, if neither retentional temporal order nor retained past data can be objectified except through the subsequent act of remembering, the retentional form of time consciousness seems to be of no use inside the phenomenological structure of experience. Someone could argue that the phenomenological theory of time has its virtue in the claim of extended intuitiveness by means of retention, which contradicts the future oriented character of object intentionality. All we can maintain in the light of phenomenological analysis is that the retentional form has its working within the whole structure of intentionality only as the site where awakening/unification/affection takes place.

Second, Husserl himself tries in an unpublished manuscript to grasp the temporal formality in terms of material fusion: "*Hyle*tic original streaming, the original impression in its stable change . . ., in which 1) the originary impression changes itself into a new impression, so that the older "yields its place"; 2) in 'yielding its place' the originary impression changes itself into retentional, while a whole coincidence in its content takes place. Regarding the first change . . ., here is also a coincidence, – the impressional is a field or a unity through the fusion of impressional simultaneousness, which has its identical, stable form" (C3 VI = C3/ 74a). Clearly, the first change concerns transition from T1 to T2 in the diagram, and the second means the diagonal line from T2 to T1. Husserl is trying to clarify temporal formation by means of material synthesis. His intention emerges more clearly in the following passage of the same manuscript: "The 'result', namely a change into something other, new, is possible only as material fusion. It is to be understood in the way that the transition of an originary impression into [another] originary impression means in reality that the new unites itself with the direct retentional change simultaneously and this simultaneous unification changes itself retentionally once more, and so on. The simultaneous unification is possible only as material fusion, . . ." (C3/75a-b). The first citation will clarify the way in which a form 'stable now' is brought about by the passive occurrence of coincidence, while the second concerns the difference between the now and the direct past. Performed consistently, this analysis would lead to the view that material fusion takes place as a passive unification and awakening without affecting the ego.

But, I think that this analysis must finally remain insufficient to do justice to the temporal diffusion insofar as it insists on the model of association as the principle of unification. It shows only the manner in which the temporal *form* emerges in each instance, but it contains no indication of how the substance of time, namely its streaming or constant transformation, sets to work. Husserl writes: "This happening of primordial streaming is not a dead happening, but the performance of the Ego is its internal motor" (Ms. C 10/15a-b, cited from Held, 1966). He immediately denies this view however by saying: "Activity in general has its "presuppositions," conditions of its possibility, which do not themselves have their source in activity (Ms. C2/66, cited from Held, 1966, 103), and no further solution can be found in his written texts which are available to us. It is Merleau-Ponty who has clarified this enigma of streaming time.

To be sure, time is a one dimensional order of 'one after another'. But, it differs from that of alphabetical order or of cardinal number, because the items cannot coexist at the same time in streaming time. The objective world does not contain such incompleteness, so that the temporal positions, which the things occupy there, are eternally fixed once determined (Merleau-Ponty, 471). Objective things (or world) are thus for us the transcendent pole of repulsion (ibid., 374), while they cannot exist but in the correlation between things and our existence (ibid., 369–370). This ambiguity of 'negation of body by things' and 'correlation of things and body' (ibid., 376) owes its possibility in the structure of time: Because it is impossible to scan all the things or all the moments simultaneously, but nothing can exist outside correlation with experience, every moment can and must be present once and only once. Thus, each moment appears one after another excluding the simultaneous presence of others, which makes time flow. The world and time has its transcendence and opacity, because each thing or presence can appear only when others hide themselves. Merleau-Ponty writes: "There exists time for me, . . . only because it is impossible for all the beings to be given bodily at once" (ibid., 484). Time must have the character of movement lest the world should lack its being in itself as transcendent.

This insight of Merleau-Ponty shows that streaming time arises only in the appearing and hiding of world. The commitment of the ego is certainly indispensable, but the entanglement of world and time, presence and absence as a whole, lies far deeper beyond the single activity of the ego in each time. It is not surprising, therefore, that Husserl fell into an apparently paradoxical situation when he tried to interpret the whole situation solely with the schema of act-intentionality. Besides, we can capture streaming time only by taking the position of 'inherence' in our experience, where the standpoint of description coincides with the described, as Merleau-Ponty wants us to accept (PP, 397 etc.).

Taking this inherent point of view, the whole phenomenon of association shows different aspects, too. Consider the case in which awakening between a black spot on a white paper in T1 and those in T2 takes place. Taking for granted the fact that sensory units owe their existence to contrast or a figure/ground structure, we can describe the phenomenon in the following fashion. It may appear at first sight that when there is a figure/ground relationship of black on white in T1 and T2, the resemblance between the two releases awakening, which affects the ego at the same time, and so on. This is a description from the point of view different than that of inherence. Because unification in both T1 and T2 presupposes the transmission of awakening, it is after awakening took place that the sensory units of black on white in T2 and T1 are formed. Seen from the inherent point of view, for the ego living the phenomena, it is not the two figure/ground structures in both temporal loci that exist before everything, but awakening, from which the difference between awakening and awakened or temporal form/content is subsequently abstracted. Consider the case where an unfamiliar face of a young woman in a crowd catches your attention. Afterwards you recognize that she resembles your aunt who died a long time ago. It was the awakening between her and your aunt which made her stand out in the crowd, reminding you of your aunt; it would be nonsensical to assume that you were already thinking of her. The stubborn dualism of temporal form/content was nothing other than a schema created by reflection in distance, which is performed after a phenomena of awakening has already happened.

Here, we must notice the following three points: First, the structure of awakening which brings about the figure/ground structure in different temporal phases leaves no room for talking about something like a sensory atom as given. We must rather say that it is a multiplied structure or 'a structure of structures' which is brought about through awakening.

Second, if awakening emerges already before the terms of awakening are articulated, this would mean that the structure of '*différance*', as Derrida puts it, is at work here. *Différance* or *suppléance* is a structure in which "a possibility brings about something in retard, to which it [= the possibility] is said to be added" (VP, 99). While we ordinarily think that a self precedes the possibility of being-for-itself in self-presentation, it is this possibility of being- or standing-for-itself which stands proxy for the self in reality. By the same token, awakening or association precedes such items like temporal form, content, and so on, to which awakening is normally said to be added. This structure looks certainly mysterious for an observer who stands outside the occurrence, but never for the eyes which take the inherent standpoint.

In the *Logical Investigations*, association was regarded as the origin of the indicative sign, as I have mentioned. In introducing it, Husserl has also imported the intrinsic features of accidental signs, which turn out to be the passivity and unreliability of association and remembering. It may be condemned as the overturning of voluntaristic phenomenology, to be sure, but never of phenomenology *per se* , because the working of association can be captured and clarified only by the inherent eyes of the phenomenologist as an occurrence within the phenomenological structure.

Third, it makes pretty good sense to separate the analysis of temporal constitution and that of material synthesis, as Husserl has *de facto* done, because a concrete phenomenon of awakening, seen in retard, implies two opposite tensions as its essential ingredients: Awakening is unification in diffusion, as well as diffusion despite unification. Diffusion and unification, both of which make up the sufficient condition of awakening, stand in the relation of "ambiguity" in Merleau-Ponty's sense. While 'contradiction' stands for the incompatibility of two opposite determinations about one and the same subject, 'ambiguity' means that two exclusive determinations are compatible in virtue of one and the same structure which generates them (Merleau-Ponty, 383). According to Sartre, our being in a situation and our being free contradict each other; they relate to each other in an ambiguous way for Merleau-Ponty by virtue of the structure of our existence which consists of projection and involvement: Our freedom is realized only thanks to involvement in a situation, as the example of a partisan bearing torture indicates, since a situation cannot work as situation without our engagement in it (PP. 513–519). By the same token, one and the same structure of awakening enables the ambiguity of diffusion and unification. All Husserl has done was to analyze diffusion in terms of time consciousness and unification by means of association.

Now, it is a good time to review the whole course of our argument. Even if we have succeeded in showing the way out of form/content dualism, a problem remains: How can these considerations contribute to solving the original difficulty arising from dualism, namely the question concerning the epistemological function of sensory *hyle* in perception?

III. Reversal of cartesian scheme

We must admit that the solution of dualism in terms of the trinity of unification/affection/awakening does not help answer the question *directly*, simply because the *Analysen zur passiven Synthesis* does not go beyond the constitution of past and of the stream of experience as the 'first transcendence' (Hua XI, 204). Far more important, rather, is to notice its contribution to helping intentional analysis transform its whole framework, which in turn will shed light on the validity of the question concerning the workings of sensory data.

The question which led Husserl to the analysis of association concerned the status of the momentary, immanent given. Having admitted the "indubitable, the unannullable validity" (Hua XI, 110) of the immanent given at the moment when it is given, Husserl declares immediately that this momentary given is insufficient or rather useless for the perceived to be a being in itself. He writes: "But what good is it [= doubtless, indefeasible validity of the present perception]?": "But the being that we grasp there is only meant as being in itself when we not only take it as a momentary datum in the mode of the present, but also as the identical *dabile* that could be given in arbitrarily iterated rememberings . . ." (Hua XI, 110). An immanent given is identified as such only by means of remembering, which in turn is enabled only by associative awakening. This was the course of argument which introduced the analysis of association. We must appreciate thoroughly the explosive implication of the sentence cited above.

Remember that in *Ideas I* Husserl has declared the absoluteness of the immanent given in the light of the structural difference of perceiving. While a transcendent perception can turn out to be false, because its object can only be given through numbers of adumbration, an immanent perception is absolute and uncancellable, because it implies no horizon that points beyond the actual given (Hua III, 88 etc.). This assumption was the most important ground leading to the Cartesian consequence, which claims the accidental positing of world *versus* the necessary indubitability of consciousness. Once admitted, it leads us quite easily to ask questions about the coincidence of the noema with the thing itself, about the origin of sensory data which must be given directly from the things, or alternatively to undertake a program to reconstruct the outer world on the basis of immanent structure.

As against this argument, Husserl introduces another aspect concerning the 'meaning' of being an object in *Analysen zur passiven Synthesis*; this leads him to the statement in question. One of the main failures in interpreting Husserl lies in the tendency to limit observation to the sphere of sensory perception, which leads to the notion of object as 'real existent' or 'intuited being'. In talking about the 'big bang', a 'cherub' or 'number', however, we surely have them as the 'objects' of discourse, but never regard them as something really existent or sensually intuited (cf. Frege, §89). The notion of object that can be applied in all these cases is designated by Husserl as 'something identifiable'. If we say 'a cherub has rainbow colored wings', we are talking about something that is identifiable as a cherub as well as having such and such wings, where it is no use asking about its existence or its intuitiveness. Husserl writes: "Being an object consists exclusively in the evidence that it can be recognized repeatedly as identical" (XXVI, 290–291), or, "An object holds true at all times . . . as the identical unifying point for predication" (op. cit., 72, cf. Hua XI, 203, Hua I, 95–96, Hua XVII, 164, Tugendhat, 234). In the *Analysen zur passiven Synthesis*, he applies this notion to the correlates of immanent perception, leading to the consequence that even an uncancellable, indefeasible momentary given does not deserve to be called an 'object' without being able to be identified repeatedly. Some would argue that Husserl, or at least my interpretation of him, commits the failure of metabasis from predicative judgment to the pre-predicative sphere. I think, however, that his maneuver was quite sound regarding the fact that the descriptive content of phenomenological analysis can and must be conveyed only by means of language. Note here that the structure of *différance* figures again, because identification which does not occur at the moment of the given is the presupposition for a momentary, present given to be called an object.

The possibility of repeated identification in turn can be secured by virtue of remembering through association. A problem arises: Not only is the process of association itself imbued by the essential accidentalness of its occurrence, but also a memory is itself far from being certain or trustworthy, as I have already mentioned. Even if we were successful in proving the advantage of the immanent domain in *general* against the transcendent world, we could not assert of *any single, individual* immanent given to be absolutely certain, which makes the former claim simply without substance.

Notwithstanding this, it is not true to the facts to condemn Husserl as an ultra-skeptic. This we can make clear by turning to an evaluation of the counterpart of the immanent given: the concept of world. His concept of 'world' is highly ambiguous. We can point out at least four different notions of it: 'horizon', 'footing [*Boden*]', 'sedimentation' and 'the sum-total of things'. Each of theses various 'aspects' is firmly embedded within the very structure of intentional correlation, which we must now examine.

At first sight, transcendent experience seems as much lacking final certainty as does inner perception. Consider the case in which you go out to observe birds and find something brown in a bush near a small river. If you suspect it to be a rare kind of bird for this region, you will continue to observe it, until you whisper to your companion: 'Look! It's a bittern!'. Husserl has defined self-evidence in *Ideas I* as the "unity of a rational positing and that which essentially motivates it" (Hua III, 316). Rational positing here means such a positing which claims to have a legitimating ground (Hua III, §136, cf. Tugendhat, 43). "The noematic proposition which is fulfilled" (ibid.) in an original manner motivates a rational positing. In the case of our bird-watching excursion, you have posited rationally that a bittern is there, being motivated by the fulfilled noematic proposition that 'a middle-size brown bird with such and such beak in a water-front bush is a bittern'. No rational positing, however, can be more than a pretense or ostentation to be the truth. Certainly, having judged something to be a bittern on the basis of visual information, you can expect or even predict certain features of its quacking or its way of feeding, anatomical data, and so on. A rational positing implies an infinite number of predicates or meaning intentions, which are prescribed by inner and outer horizon as the rules for obtaining adequate determination of the object in question (Hua III, 330). A rational positing as a whole, including horizonal intentions, most of which are not fulfilled yet, is posited as certainty. But, it can lead to disappointment, because we cannot exclude the possibility that a fact contradicting the horizon of bittern will turn out to be the case. If you realize that the object cannot move at all, you will say: 'It was not a bittern, but a decoy'. By contrast, something can be said to exist truly, if and only if the possibility of disappointment is completely excluded. Husserl writes: " 'Truly existent object' is equipollent to the object rationally posited in an original and perfect mode" (Hua III, 329), or "The eidos of truly existent is equivalent to the eidos adequately given and evidently posited" (ibid., 332). Such a state of affairs is realized only when every possible component of the horizon is fulfilled without leaving any open element. However, this is in principle impossible in the light of infinite numbers of possible predicates. For Husserl truth or true existence must remain the "ideal in the Kantian sense." Note that this is not only impossible to be realized in the actual course of active cognition, but also unable to be seen or intuited by phenomenological eyes as something *positive* because its status in the phenomenological structure is ascertained only by observing its *negative* working in experience. All we can confirm positively is that no truths at hand are absolute, but they are mere defeasible ones in the light of Kantian ideal (Nuki, forthcoming).

This unavailability of final certainty does not hold for the various modes of the world. 'Horizon' is the indispensable component of intentional correlation: It prescribes the way that the adequate given is pursued in order to confirm a rational positing. 'The sum of things' would be the whole of things legitimated as truly existent, which is actually unrealizable, but works as an ideal for scientific research. Because active cognition cannot take place *ex nihilo*, Husserl writes in *Experience and Judgment*: "Before active cognition sets in, various objects are always already existent for us; they are given in the mode of certainty *simpliciter*" (EU, 23), or "The surroundings [of active cognition] is co-existent as the sphere of pre-given. This is a passive pre-given, which always already exists without activity or any directing of the grasping gaze. This sphere of the passive pre-given is presupposed by every kind of active cognition. . . . To put it in another way, every kind of cognition is preceded by the world at each instance as a universal *footing*" (EU, 24, italic mine). On a bird-watching excursion, the bird is the only object we have, while its surrounding, such as bush, road, river or the sky remain the completely un-objectified footing. However, they motivate our positing of the bird as a bittern, because if spotting something brown moving in a bush, not near a river, but in the desert, we would posit a completely different thing, e.g. a kind of reptile. The surroundings must be given to us previously, and only in a passive manner because they are incapable of being examined by our active cognition. If every single detail of the surroundings must be examined beforehand, our active observations would be simply impossible, or even worse, we must fall into infinite regress: an activity of examining a surrounding presupposes another surrounding, which must again be examined in advance, and so on. This certainty *simpliciter* of footing has no grounding, because active cognition alone can give something a legitimating foundation. Lastly, 'sedimentation' can be regarded as a kind of footing which is seen from the context of history or tradition.

All the aspects of world are thus intertwined in the structure of intentional correlation, in motivating active cognition, guiding it and being pursued in it. It is not only needless, but also pointless to 'demonstrate' the existence of world anew which is assumed to be 'outside' intentional correlation. Even terminology such as "being in the world" is dangerous insofar as it can be interpreted as a relationship between two ready-made entities, thus failing to grasp the connateness or co-radicalness of transcendental subjectivity and world.

In sum, the ripest theory of Husserl has shown up as the reversal of Cartesian scheme: In *the latter*, the absoluteness of inside was pretended to be a bridgehead to reconstruct the accidental, outer world. *In the former*, 'inside' is declared to be dependent on the accidental happening of association and is deprived of its absoluteness, while simple reversion to naive naturalism is averted by establishing the connateness of subjective activity and objective world in the same structure of intentionality. This is the rough sketch of consequence which is lead from the phenomenological theory of association. The question arises again: How could it contribute to solving the difficulty of form/content dualism?

IV. The status of hyle

I think that the puzzle over the function of the sensory datum has its psychological background in the assumption that *hyle* is the only access to, as well as the only immediate given from, the things themselves existing *beyond* the structure of representation or intentionality. This assumption is simply false in the light of intentional analysis. First, a *hyletic* datum alone cannot be the decisive proof or criterion for existence of a certain perceptive thing, because one and the same sense datum can be interpreted in various ways (Hua III, 230). Second, the supposition that a sensory datum can be the decisive *criterion* for the existence of a thing leads to the view that a proposition concerning the existence of a thing can be confirmed because it is true, which commits the failure of "theory from above," as Husserl calls it (Hua III, 46, Hua XVII, 286, 284, Hua I. 34). Husserl holds a reversed view that propositions can be true only because they are confirmed, for it is only the experience of confirmation, but not the supposed truth itself or its relationship to the purported criterion, that we can approach. We cannot even tacitly assume the thing itself beyond our experience as well as the truth without taking into account the possibility of its confirmation. Such a supposition commits the failure of "sign theory" or "picture theory," which Husserl criticizes in *Ideas I*, and, for a phenomenologist promoting the inherent point of view, it makes no sense to take a "God's-Eyes-View," which compares our experience and the thing, itself beyond it from a standpoint hovering over them. Of course, phenomenology will not endeavor to integrate everything inside subjectivity or its intentional content. Husserl admits willingly a horizonal indication beyond the actually given or even the intended. This leads him to appeal to the Kantian ideal (Nuki, forthcoming). The only way to speak phenomenologically about 'Outside' is to keep its place lying not *behind* the current given, but *ahead* for the future-oriented activity; this leaves no room for us to ask the epistemological role of the sensory given.

V. Conclusion

The structure of our experience or our way of being as a whole shows up as a texture of the ruptured, ambiguous structures of connateness folding each other from the lowest level of awakening to the higher dimension of intentional correlation. Here, each component supports the other by virtue of its being placed in a systematic whole; it contains no decisive instance such as the Cartesian ego or sensory datum, much less the objective world: The objective world loses its overwhelming certainty in the light of the Cartesian reduction, while the theory of association robs immanence of its absoluteness.

It is at this point that we can perceive the failure that such a subtle interpreter as Derrida has committed. I can indicate only an outline of my counter-criticism. Without mentioning the problems of ideal essence (VP, 4–5, 8, 58) which Derrida attacks by ignoring that Husserl's system was always in the works, or of the Kantian ideal (PG, 39, fn., OG, 147, 150–151, VP, 8), of which he will not notice the way of appearing in hiding itself, it is altogether misleading or a misreading to say that everything demonstrated phenomenologically is supported by the 'purported' presentation of the instant or the punctuality of now (VP, 67) where he finds the residual of indicative sign or *différance*, which proves the breakdown of Husserl's endeavor to reduce everything to the immediate given. Derrida's claim does not do justice to the fact that Husserl's system as a whole has neither any absolute instance, from where everything might be clarified or even grounded, nor any kind of all-dominating scheme, with which every mechanism should be grasped, such as apprehension-model, form/content dualism or even *différance*. It is also beside the point to claim that Husserl's system has been built with the methodological principle of intuition (OG, 151, VP, 3, 67). As we have already seen, Husserl's entire theory was worked out with a number of arguments and models, which, it is true, should be tested by intuition, but which enlarge and enrich the range and resolution of intuition as well. Remember that for Husserl an intuition consists of signifying intention and its fulfillment: You cannot recognize a bittern, even if it stands right before you unless you had knowledge of it beforehand such as you would find in a bird atlas, which I suspect to be the case by almost every reader of this paper.

All Derrida has overlooked is the fact that Husserl's system is something like a complex full of capillary blood vessels without any carotid artery. The carotid artery fosters everything, but a successful attack on it would prove fatal. That his system lacks such a vital spot is exactly the strength of phenomenology, which might be overlooked perhaps by the historical Husserl, but which is surely opened up by his radical overturning of stable instances from the inherent point of view.

Acknowledgement

I would like to thank Dr. Felix O'Murchadha for his kind help in the linguistic formulation of this paper. I also appreciated the kind advice and criticism of Dr. Christian Wenzel, Professor Tetsuya Sakakibara and again Dr. O'Murchadha. I am grateful to Dr. Lohmer, the director of the Husserl-Archives in Cologne who facilitated me in my research. I also wish to thank very much Professor Bernet, the director of Husserl-Archives in Leuven, for his help in allowing me to cite the unpublished manuscripts of Husserl. Last

290

but not least, I am heartfully grateful to Professor Dr. Klaus Held, who has not only given the most helpful advice for this paper, but also enabled me to stay in Wuppertal for one year as an Alexander von Humboldt researcher. This paper was written as the result of the research supported by the foundation.

References

Brand, G., *Welt, Ich und Zeit*. Martinus Nijhoff, 1969.

Derrida, J., *Edmund Husserl, L'origine de la géometrie*, traduction et introduction, PUF, 1962 (OG).

Derrida, J., *Le probléme de la genése dans la philosophie de Husserl*, PUF, 1990 (PG).

Frege, G., *Die Grundlagen der Arithmetik (The Foundations of Arithmetic)*, Oxford and New York, 1953.

Fink, E., *Nähe und Distanz*, Alber, 1976.

Held, K., *Lebendige Gegenwart*, Martinus Nijhoff, 1966.

Held, K., *La voix et le phénoméne*, PUF, 1967 (VP).

Held, K., *Phänomenologie der Lebenswelt, Ausgewählte Texte II*, Reklam, 1986.

Holenstein, E., *Phänomenologie der Assoziation*, Martinus Nijhoff, 1972.

Husserl, E., (Unpublished manuscripts) C3VI = C3, C10, C21.

Husserl, E., *Analysen zur passiven Synthesis*, Martinus Nijhoff, 1966 (Hua. XI).

Husserl, E., *Zur Phänomenologie des inneren Zeitbewußtseins*, Martinus Nijhoff, 1966 (Hua X).

Husserl, E., *Logische Utersuchungen*, Bd. II/ 1, 1900/0 1, Max Niemeyer, 1968.

Husserl, E., *Erfahrung und Urteil*, Felix Meiner, 1972 (EU).

Husserl, E., *Cartesianische Meditationen*, Martinus Nijhoff, 1973 (Hua. I).

Husserl, E., *Formale und transzendentale Logik*, Martinus Nijhoff, 1974 (Hua. XVII).

Husserl, E., *Ideen zu einer reinen Phänomenologie und phänomenologischen Philosophie*, Martinus Nijhoff, 1976 (Hua. III).

Husserl, E., *Vorlesung über Bedeutungslehre*, Martinus Nijhoff, 1987 (Hua. XXVI).

Husserl, E., *Die Krisis der europäischen Wissenschaften und die transzendentale Phänomenologie, Ergänzungsband*, Martinus Nijhoff, 1993 (Hua. XXIX).

Merleau-Ponty, M., *Phénoménologie de la perception*, Gallimard, 1945.

Nuki, S., 'Das Problem des Todes bei Husserl' in *Phänomenologie der Praxis*, Königshausen und Neumann, 1989.

Nuki, S., 'Phenomenology as Calculus?' in *Japanese and American Phenomenology*, Kluwer Academic Publishers, forthcoming.

Tugendhat, E., *Wahrheitsbegriff bei Husserl und Heidegger*, Walther de Gruyter, 1970.

Welton, D., 'Structure and Genesis in Husserl's Phenomenology' in *Husserl Expositions and Appraisals*, University of Notre Dame Press, 1977.

Yamaguchi, I., *Passive Synthesis und Intersubjektivität bei E. Husserl*, Martinus Nijhoff, 1978.

The translations cited here of Husserliana XI, *Analysen zur passiven Synthesis*, follow the forthcoming English edition, Edmund Husserl, *Analyses Concerning Passive and Active Synthesis: Lectures on Transcendental Logic*, trans., Anthony J. Steinbock, Husserliana Collected Works (Boston: Kluwer Academic Publishers).

Note about the author

Shigeto Nuki (1956) studied philosophy at the University of Tokyo. He is Associate Professor of Philosophy at Saitama University. From 1986–1987 he was a research fellow of DAAD in Wuppertal, Germany, and from 1996–1997 he held an Alexander von Humboldt scholarship in Wuppertal. His works include "Das Problem des Todes bei Husserl. Ein Aspekt zum Problem des Zusammenhangs zwischen Intersubjectivität und Zeitlichkeit," in *Phänomenologie der Praxis im Dialog zwischen Japan und Westen* (1987), "On the Self-Sufficiency of Language" (in Japanese, 1991), and "Language as Culculus" in *Phenomenology: Japanese and American Perspectives*, edited by Burt C Hopkins, Kluwer Academic Publishers (forthcoming). He is also the translator of J.N. Mohanty's *Husserl and Frege*. His current interests include the phenomenology of temporality, logical semantics, and the theory of narratives.

Continental Philosophy Review **31**: 293–305, 1998.
© 1998 *Kluwer Academic Publishers.*

Colors in the life-world

JUNICHI MURATA
*Department of History and Philosophy of Science, The University of Tokyo, 3-8-1 Komaba,
Meguro-ku, Tokyo 153-0041, Japan*

Do things look red, because they are red? Or are things red, because they look
red? Naive realists would answer positively to the first question, and idealists
positively to the second. But since Galileo natural scientists have provided
a more radical answer: If there were no human beings, there would be no
colors on the earth. To be exact, there are no colors in the objective world,
and things in the world have no color. Colors are only subjective phenomena,
like "hallucinations."

Husserl has taken a very clear stance against this "scientific realism" concerning so called "secondary" qualities or "qualia," such as colors, sounds,
and so on. Colors are "sensory qualities": they are not to be confused with
"sense data," but are to be taken as "properties of objects which are really
perceived in these properties" (*Krisis*, p. 28). Therefore, according to Husserl,
there are colors in the world, at least in the lifeworld. But *how* do they exist?
This is the question that I would like to take up in this paper.

I. Mixed color

There is a famous phenomenon in which two things seem to have the same
color, although they reflect totally different spectral components of light. It
is called "metamerism." Metamerism is a phenomenon that color scientists
and philosophers use in order to "demonstrate" that colors are subjective
phenomena. But is this a necessary consequence?

It is thoroughly trivial that two things may show the same form when
we see them from a certain direction even though they have different spatial
structures. The spatial form has various "sides" or "aspects," and each time
we see it, it shows one corresponding aspect. Husserl has called this structure
of phenomena "adumbration" [*Abschattung*], and saw the same structure in

the sensory qualities of things. Color is therefore considered to "adumbrate" itself. But does that mean that color has various "aspects"?

Let's take a familiar example of mixed color. The picture of a color television consists of three kinds of luminous colored points: red, green, blue. When we see, for example, the yellow color of a fresh lemon in a picture, components of light reflected from it are mostly made of red and green spectral light and are very different from those reflected from a real lemon. The yellow color of the television picture is not pure spectral color but mixed color. Psychologists often describe this situation by saying that the mixed color in the television picture is an *illusion* and that the color television uses this kind of illusion.

Why must our perception of the yellow in the television picture be taken as an illusion? Because, answer the psychologists, there are *really* only two kinds of luminous colored points, i.e. red points and green points, in the place where the yellow color is seen. These luminous points, however, cannot be identified, for they are too small to see.

But this argumentation lacks necessity: If we accept this kind of reasoning as an explanation of an illusion, we must regard almost all of our perceptions as illusions. For example, when we see a picture of Monet and see water lilies in it, our perception of water lilies is an illusion, because there are really only colored spots on the canvas. Or although tap water looks transparent, the appearance of the tap water is an illusion, because it really contains many kinds of things that we can identify through a microscope.

Against this argumentation we can describe the situation in another way, using the Husserlian concept of "adumbration."

First, there is no reason why the appearance of the color seen through a microscope must be taken as "real" or "true," whereas the appearance of color seen with the naked eye should be taken as illusory. If we should use a microscope with more magnifying power, the appearance will change even more. The appearance of a color through a microscope is just one mode of appearance along with various other modes, and *as a mode of appearance* it has no reason to be taken as having a privileged status. Every appearance must be taken as having equal status. In this sense, color has various "aspects," just like a spatial form.

Second, if the magnifying power of a microscope is so high that the scale of objects and that of wave lengths of light do not differ appreciably, then the objects will be unable to reflect light and not be seen. In this case, the contrast will no longer be that between the real color and illusory color, but a contrast between something colored and something colorless, in other words, the contrast between the visible and the invisible. And in consequence, all the perceptions of colors are to be taken as illusory because what "really" is must be considered as belonging to the realm of the invisible. In fact, this is the direction of argument that the Modern scientific view since Galileo has taken. Could we also use the concept of "adumbration" against this kind of argumentation of "scientific realism"?

Perhaps this can be presented in the following way: Our visible world is constituted not only of visible components but also of invisible components. Color has visible "aspects" and invisible "aspects." Both "aspects" are inseparable. If we can describe the phenomenon in this way, we can defend our perceptions of colors in the lifeworld against the eliminalistic monopoly of scientific realism, without falling into the reverse extreme position, that is, the position that makes our lifeworld something absolute and incorrigible. In this way we come to a position that could be classified as a kind of "double aspect theory" of the mind-body problem without taking the two "aspects" as autonomous and closed. While the invisible scientific world shows only one inseparable aspect of the lifeworld, the visible colorful world cannot be closed and is always open to a variety of scientific research. How these aspects are related to each other has long constituted the core of the mind-body problem. Before we go into this problem, we would like to remain in the visible world a little longer.

II. Incommensurability of colors

In the first section I have described the problem as if we could perceive color independently from other factors, and we must correct this now.

Color shows itself, firstly, always only as the color *of* something and secondly as the color of something *in a certain situation.*

When we see a yellow color in a television picture, we see not only the yellow in general but the yellow *of a lemon.* If we take this point into consideration, the red and green spotted surface, which appears through a microscope, is unnatural and abnormal. If we see a red and green spotted lemon, either we think that the appearance is abnormal or we take the lemon itself to be abnormal. In any case, we presuppose that there is a normal mode of appearance which is taken to be a criterion we can then use to evaluate the appearances of a color.

If we nevertheless have not taken this trivial point into consideration in the above example, the reason must be that the red and green spotted surface, which we see through a microscope, has not been seen as the surface of a lemon but as a surface of the cathode-ray tube (CRT) of a television. The object of the perception itself has changed. But so long as colors are always colors of some object and so long as this object takes a role in determining the criterion of the appearances of colors through which the object is perceived, and so long as the object is changed in the above case, colors in both cases stand under different criteria and therefore cannot be compared directly. The red and the green of the color spots on a television picture and the yellow of a lemon are "different kinds" of colors.

To every color of an object corresponds a standard situation, in which the appearance of the color is taken as optimal and in this sense as "real." In our everyday life we presuppose this "secondary objectification" (*Ideen I*, p. 82) in some way or other. That is why when we want to buy clothes we try to see them not under artificial light but under the sun. Usually we consider the day light of the sun to be standard lighting. But on the other hand, we cannot ignore that this everyday criterion has only a limited validity. For example, what kind of color does the moon have under this "standard lighting"? What about the color of the setting sun or the color of a neon sign?

What is the "standard situation" in which the color of the moon is determined? Is it the night in which the moon shines in the sky? Or is it the situation in which we view the moon from a spaceship? Or is it the situation in which we find ourselves on the moon? If we take such different situations into consideration, it is difficult to determine one universally valid criterion that applies to all of them. We must rather determine a standard condition corresponding to each situation. That means the criterion of colors, according to which the kind of color the object has is determined, changes with each situation. Even if the object remains the same, as in the case of the moon, colors which appear in the various and "incompossible" situations do not stand under the same criterion and cannot be compared simply with each other. The moon shows its "incompossible" colors which correspond to "incompossible" situations. Perhaps we could say that not only colors but also the "being" of the moon itself is different and incompossible corresponding to each situation, so long as the way of being perceived is concerned. Merleau-Ponty has pointed out this peculiar structure of the perceptual world. "But for me the perceiver, the object a hundred yards away is not real and present in the sense in which it is at ten yards ..." (Merleau-Ponty, 1945, p. 348; tr. p. 302).

In our perceptions of our everyday life we are usually not conscious of such a difference, and cross the border between incompossible situations very easily. Nevertheless our perceptual world is full of possibilities for being surprised with "other kinds" or "new kinds" of colors that we encounter through various experiences, although it is seldom to be so surprising as in the case of travel from the earth to the moon. The identity of an object, which is constituted through perceptual processes, can establish itself only through such adventurous experiences. If we nevertheless regard these various appearances as having equal status from the beginning and with Husserl as various "adumbrations" of an "objective identity," then we presuppose with it that the first "idealization" [*Idealisierung*] has already been accomplished. The Husserlian thesis that colors are properties of perceived objects is situated in this level of the "constitution of the lifeworld."

While up to now we have mostly taken into consideration only a *successive* "incompossibility" of perception, there are *simultaneous* "incompossibilities." That is, it is not rare that we experience the various situations that we have covered in the above example in one perceptual field simultaneously.

Can we say that the yellow of the moon which shines in the sky is the same as the yellow of a lemon which is lighted by a lamp? So long as there is no common "standard condition," it is rather adequate to say that the place where the moon lies and the place where the lemon lies are different or different kinds of places, and that the heavenly world and the earthly world are "different" worlds. We can find an impressive description of this remarkable structure of the perceptual world once again in Merleau-Ponty.

We must not neglect to refer back to the important remarks of psychologists who have shown that in free and spontaneous perception objects standing in depth do not have a definitive apparent size. The objects which are in the distance are not larger than the rules of perspective indicate, and the moon on the horizon is not 'larger' than the piece of a franc which I have in my hand, or at least it has not a largeness which would become a measure of two objects. The moon is a 'large object in the distance'. The largeness which matters here is like warmness or coldness, a property which adheres to the moon and which cannot be measured by a certain number of aliquot parts of the piece of money. (Merleau-Ponty, 1969, p. 72).

That an object appears in depth in a perceptual field means that it shows a "largeness" which has no common measure with other objects, and in this sense is "incommensurable." Painters have long been conscious of this incommensurable depth structure of our perceptual world. For example, they have developed a technique to represent the essential difference between the foreground and the background. "Both the Graeco-Roman and the Chinese masters of the genre refused to acknowledge the middle ground and preferred to veil it in haze or mist. Thus they achieve a sharp distinction between houses or rocks in the proximity which are made to look solid, and the mountains or trees in the distance which are projected as flat shapes against the sky" (Gombrich, p. 87). This technique could be interpreted as the method painters used to make the difference of the "being" of things in depth perceptible.

We can find the same depth structure in the phenomena of colors as well. So long as colors such as the yellow of the moon and that of a lemon, which are situated in different places in a perceptual field, cannot be compared directly with each other and are in this sense incommensurable, we can also say that they appear in the "depth" related to colors. Color has not only "aspects" but also "depth." With this characterization of the depth of colors we can now understand the significance and the consequence of the concept of "adumbration" [*Abschattung*,] with which Husserl has characterized not only the perception of form but also color. Only something that appears in depth can have "aspects" in the true sense, because without depth there can be no "aspects" nor "sides," but only parts.

Color in "free and spontaneous perception" cannot be considered as something that can be identified with a functional relation of elements in a commensurable dimension. Nevertheless, Newton began trying to breed the "raw and wild" being of colors and to make them manipulable: He made the "free" light of various colors pass through a slit into a dark room, made a thin ray of light, resolved it into the spectral lights through a prism, and then re-mixed these resolved lights, and so on. The science of color, the founder of which was Newton, could be seen as a "forcible" attempt (to the same extent that the perspectivism which began in the Renaissance could be seen as a "forcible" attempt), to make incommensurable spatial depth commensurable. The result of this attempt was the "discovery" of the phenomenon called "metamerism."

In the Newtonian view, colors are first abstracted and separated from both the things to which they adhere and from the situation that they inhabited, and are then analyzed, manipulated, and calculated. The yellow of the lemon and the yellow of the CTR of the television are on the one hand considered to be the same and on the other hand considered to be totally different with regard to the constituents of spectral colors. In this way the first constitution in the lifeworld is skipped, and the phenomenon of color is decomposed into one thing purely psychological and another thing purely physical. The simultaneity of the incompossibles and the incommensurables is resolved to the simultaneity of the compossibles and the commensurables. But through that process emerges another paradox of the simultaneity of the incompossibles, that is the simultaneity of the visible and the invisible.

III. "Double aspects" of colors

"The ray has no colors"

"The ray has no colors." This thesis, which Newton proposed in his "Optics," has become now a slogan of color scientists who seem to think that only after this thesis has been posited can colors become an object of scientific research. But if the ray of light has no colors it cannot be seen since things that have no colors are essentially impossible to see. How can one then explain the fact that we see the rays of light and their colors?

Newton's answer to this question is similar to Locke's. While the ray of light has no colors, it has a "power" or disposition to make colors. The red ray must therefore be called the "red-making-ray." But this kind of answer cannot solve the difficulty; it simply transfers it to another problem, that is, the problem of explaining how colors can be "made" from colorless rays. The answer would be: through the process in which the ray is received by a perceiving subject.

But how and where can the invisible ray be transformed to a visible and colored ray when the ray is received by a subject? Is there any transformation process in the eye, in the nervous system, or in the brain? Husserl has already very clearly pointed out the absurdity of this way of thinking.

> Not even a Divine physics can make simply intuited determinations out of these categorical determinations of realities which are produced by thinking, any more than a Divine omnipotence can bring it to pass that someone paints elliptic functions or plays them on a violin. (*Ideen I*, p. 102; tr. p. 123)

If we want to avoid this "mythological" (*Ideen I*, p. 101) process, which is impossible even for God, there seems to remain only two alternatives. One is the eliminalistic strategy, according to which the existence of colors itself must be abandoned. If no colors exist, neither is there a mythological trans-formation process, and we need not be bothered by the so-called mind-body problem.

But seen from the constitutional point of view, I think this eliminalistic position is in reverse order. It is not the case that we have first physical, chemical, and physiological knowledge about our color perceptions and only then on the ground of these kinds of knowledge do we evaluate the ontological status of colors. Rather conversely we begin scientific investigations only on the ground of our various perceptual experiences of colors. Without these experiences we cannot carry out either scientific observations or experiments related color phenomena, nor can there be research into objects, although our perceptual cognition can and must sometimes be revised through the results of scientific investigations. Husserl has coined this constitutional fallacy "the forgetfulness of lifeworld" and Whitehead "the fallacy of misplaced concreteness."

Besides, even if there should be ontologically only material processes and nothing more, and color perceptions should be seen as "hallucinations," there remains the problem of explaining why we would have developed such experiences as color "hallucinations" over time, and how it is possible that such "hallucinations" "exist" in the material world. It is in principle impossible to answer these questions in the eliminalistic view, so long as in this view colors have only redundant status.

If we avoid the constitutional fallacy and if the problem concerning the relationship between the material processes and qualitative experiences remains in any case, there remains for us only a second alternative, that is the alternative that we presuppose from the beginning the visible colorful world. This world is originally filled with incompatible and incommensurable colors, and then on the ground of the first "idealization" it is filled with various colors as properties of things. According to this phenomenological position, all objects that we encounter in the lifeworld have some color, and therefore we must say that the rays have colors. It is not that the colors emerge through the "mythological" process of the transformation from the invisible to the visible, but the reverse. The scientific determinations of the rays of light are nothing but "the experimental-logical determination of the Nature given in intuition simpliciter" (*Ideen I*, p. 101; tr. p. 122). The physical determination belongs to the higher and later level of constitution. And only on the ground of this constitutional level can we say that the theoretical determinations of science show only one "aspect" and the other "aspect" appears filled with various colors.

Phenomenological model of "Lived Body"

There remains now the difficult problem of making understandable how these two "aspects" are related with each other.

A special problem related to color is that color is not definitely determined only by the wavelength of the ray of light or with the structure of the surface of things, so that it cannot be reduced to some physical property, making the objectivistic view about color unsustainable. One of the most conspicuous examples that shows this circumstance is the color brown. Brown can neither be found in the spectral components of sun light, nor made only by mixing spectral lights. Brown is not a "film color" such as the color of spectral light or the color of sky but a "surface color," i.e. a color of an opaque object.

On the one hand, the surface of a brown thing reflects a light whose wavelength is the same as that of yellow light. On the other hand, brown is differentiated from yellow in the respect of the reflectance, meaning that it is different from yellow because of the difference of brightness. In this sense brown can be called a kind of "dark yellow," especially in contrast with surrounding things. Therefore there cannot be brown color that shines more brightly than surrounding things. Wittgenstein indicated this characteristic of brown in the following way: " 'Brown light'. Suppose someone were to suggest that a traffic light be brown" (Wittgenstein, III-65).

In this way the objectivistic approach, in which color is to be reduced to some physical property such as wavelength or surface reflectance, cannot be sustained, and other factors, for example the contrast of colors, which is a phenomenon related to the whole organizational structure of the perceptual field, must be taken into consideration. In addition to that, when it comes to the characteristics of colors, such as their "logical relation," for example, the incompossibility between red and green, or yellow and blue, we must also take various other factors into consideration. In order to explain such a "logical" relation of colors, we need to refer to the structure of the retina and also to that of the nervous system. There is also much research concerning how "illusory colors," i.e. colors that are caused by an after-image effect, occur, corresponding to the changes of the retina, of the nervous system, or of the brain.

On the ground of this knowledge, we must recognize that there is a "close" relationship between color perception and the material process, which is constituted by surface structure, rays of light, the retina, the nervous system and the brain. But this relationship cannot be a causal one, because the causal relation is only meaningful either in the material field or in the psychological field. Parallelism would be unsatisfactory. We have now other new versions, such as identity theory or functional theory. But these positions have not yet cleared themselves of the suspicion that they fall into fundamental difficulty, especially when it comes to the "qualia" problem (cf. Nagel).

Perhaps phenomenologists could propose a possible model here.

Merleau-Ponty has impressively described phenomena of the extension of the lived body through artifacts, for example, the car of a skilled driver, a feather in the hat of a lady, or a blind person's stick. What matters here is not the body as object but the body as an organ for perception and movement, which makes the perception and the movement possible, and which is not objectively cognized but "lived." If the stick of a blind person becomes a part of a lived body and "transparent," the object can be "touched" directly through the stick. On the other hand, if a part of the stick appears and is perceived in some way, the perception of the original object will be hindered and changed. Here is a remarkable structure of appearance and disappearance, or the structure of being open and being hidden. (Concerning the significance of this phenomenological model of embodiment in relation to "qualia" consciousness, see Murata 1997).

There is no fundamental difficulty in transferring this structure concerning the appearance of tactile perception to visual perception. When we take into consideration the phenomenon of contrast and constancy, which are characteristics of surface color, this structure of appearance and disappearance is especially conspicuous.

Let's take the following famous experiment concerning the constancy of colors. We prepare two boxes whose insides are painted white and black respectively, and one is dimly lighted, the other brightly. We can arrange these lightings so that we cannot differentiate the colors of these boxes, seeing only a dim gray space inside these boxes when peering inside through a small peephole. But when we put a white paper into the black box, or a black paper into the white box, the appearance changes dramatically. Immediately the former is seen as the white box lighted dimly, and the latter as the black box lighted brightly (cf. Merleau-Ponty 1945, p. 355; tr. p. 307).

This experiment shows clearly that the constancy of colors is possible only on the condition of the contrast of colors and that on the ground of this contrast the differentiation of the lighting and the being lighted occurs. According to Merleau-Ponty, an "organization of color itself" occurs here and an institution of "the structure of the lighting and the being lighted." (Merleau-Ponty, 1945, p. 355).

But why is the differentiation between the lighting and the being lighted so important for the constancy of colors? Our answer would be the following.

Through this differentiation the lighting retreats from being the object which is the focus of perception to its "background" through which we see colors. That means, the lighting becomes "invisible," in the sense that through it something colored becomes "visible." Without this structure of organization we could not see the *surface color* of things, and the constancy of color could not be retained, as the case of the above experiment shows. In the first case of the experiment we could say that the colors of the lighting is the object because our gaze is stopped at the level of the lighting and cannot go through it to reach the surface of the boxes. Only when the contrast of colors has been constituted and the differentiation between the lighting and the being lighted is established, does it become possible for us to see through the lighting and see the surface of things.

In this way we can find a structure in the perception of colors similar to the case of tactile experience. One of the important points of the phenomenological model of the "lived body" is that the relationship between the "visible" and the "invisible" is compatible with the order of constitution. As we have already seen, the physical, chemical, or physiological factors, which are conditions of color perception, are situated at the later level of constitution in contrast with the color perception itself. When we follow the phenomenological model to interpret the relationship between factors that belong to scientific research and the color that we experience directly, the former factors are considered to be cognized as objects only *indirectly* and only on the ground of color perception, while the colors are considered to be *direct* objects of perception. The factors that are "lived" through in the experience can become objects of cognition only indirectly and only later than the experience itself. If we understand the sense of the phenomenological model in this way, there is no fundamental obstacle to extend this model to other factors of perception than the lighting which we have mainly considered above.

It can then be said that when we see a certain color of an object, we see not only "through" a certain ray of light, but also "through" a certain structure of the retina, "through" a certain structure of the nervous system, and "through" the brain. To the extent that rays of light, the retina, the nervous system, and the brain become "embodied" and "transparent," the colors of an object correspondingly will be visible.

Another important characteristic of this phenomenological model concerning the relationship between the visible and the invisible is, I think, that we can *experience* the change from the visible to the invisible and vice versa as a kind of gestalt change or aspect-switch. A color can become visible insofar as the other "side" or "aspect" of it becomes invisible. And if some factor of the embodiment relation does not function in a normal way, and if the embodiment relation is not more "transparent" and becomes "visible" in some way, the appearance of the original color will change greatly or be diminished and other kinds of color appear. If there is such an essential relationship between the two "aspects" of a color, we cannot identify what a certain color is unless we grasp both of its "aspects." And it is exactly this embodiment structure or the invisible aspect that creates the "depth" of each color. In this sense, every color has its own "depth."

When what is said above is admitted, where is the place of a color?

One answer is that so long as we see a color through many factors of perception, the color is where it is seen. That means, in the case of *surface color* in the normal condition, it is on the surface of things. This characteristic of surface color constitutes the way of the existence of color. The browness of the table is considered to continue to exist when the light is turned off or when nobody sees it. In this sense the existence of the surface color is "independent" of the conditions of perceptions and of perceivers, just as the form of things has an independent existence.

But on the other hand, so long as the color has its proper "depth" and the factors that constitute the depth of color can be considered another "aspect" of the color, the place of the color is not only on the surface of an object, nor in the ray of light, nor in the retina, nor in the nervous system nor in the brain, but rather it is realized in the whole system, which comprises all of these factors together. In other words, so long as these factors can be taken as constitutional conditions on the ground of which the appearance of each color is possible, the "ontological" place of colors is considered to be the whole system, which include factors from the surface of things through rays of light to the brain, i.e. the lifeworld in which we live. These are the consequences of the Husserlian thesis that colors "adumbrate" themselves.

IV. Coexistence of lifeworlds

The above consideration of color perception was made mostly from a static point of view. But the constitution of colors occurs factually in the course of a long evolutionary history in which various perceivers of colors and various colors of things emerge and develop together. To the extent that we have embodied and adapted ourselves to a certain range of electromagnetic waves, it has become possible for us to see corresponding colors. In order to realize this, we have also had to develop and embody a certain structure of the retina, the nervous system, and the brain. The situation is not fundamentally different in the case of other living creatures. Animals and insects have developed their perceptual organs with which various colors can be identified. On the other hand, plants have developed colorful flowers and trees in order to be identified by and "use" the animals and insects. The colorful world, which we now perceive, is the result of a long process of the *co-evolution* between various living creatures and their lifeworlds.

It is well-known that bees perceive ultraviolet colors. What it might be like to experience such colors for bees is beyond our ability to understand. The colors of other living creatures are in principle incompossible with "our" colors. In this sense the colors of the other living animals are in principle invisible, and these colors cannot be considered to be other "aspects" of our visible colors in the sense of this concept we have considered up to this point. Nevertheless this does not mean that we can revive the view that the colors of the other animals are subjective sensations and do not exist in the world. Rather the same thing can be said in the case of the colors of other animals as in the case of "our" colors. The ultraviolet "color" that bees perceive is constituted by the whole system of factors and lies in the lifeworld of the bees. Besides we see and enjoy the colors of flowers, which have coevolved with perceptual organs of bees, "from our point of view" and in "our" lifeworld. In this sense although the colors of other animals are impossible to see, we could say that we see the other "aspects" of that invisible color. The colors which we see are other "aspects" of the invisible colors of the other animals, and how we perceive colors shows the form of coexistence between "our" lifeworld and the lifeworlds of other animals.

References

Husserl, E. 1950, *Ideen zu einer reinen Phänomenologie und phänomenologischen Philoso-phie*, Husserliana Bd. 3, ed., W. Biemel, Martinus Nijhoff, The Hague (citation pages from the 1st edition; abridged to *Ideen I*); translated by S. Kersten, Martinus Nijhoff, 1982.

Husserl, E., 1962, *Die Krisis der europäischen Wissenschaften und die transzendentale Phänomenologie*, Husserliana Bd. 6, ed., W. Biemel, Martinus Nijhoff, The Hague (abridged to *Krisis*).

Merleau-Ponty, M., 1945, *Phénoménologie de la perception*, Gallimard, Paris; translated by Colin Smith, Humanities Press, 1962.

Merleau-Ponty, M., 1969, *La prose du monde*, Gallimard, Paris.

Murata, J., 1997, "Consciousness and the mind-body problem" in *Cognition, Computation, and Consciousness*, ed., I. Masao, Y. Miyashita and E. Rolls, Oxford University Press.

Gombrich, E.H, 1974, "The sky is the limit: The vault of heaven and pictorial vision," R. MacLeod and H. Pick eds., *Perception, Essays in Honor of James J. Gibson*, Cornell UP, Ithaca.

Nagel, T., 1979, "What is it like to be a bat?" in *Mortal Questions*, Cambridge UP, Cambridge.

Whitehead, A., 1948, *Science and the Modem World*, Mentor Books, New York.

Wittgenstein, L., 1977, *Remarks on Color*, ed., G.E.M. Anscombe, University of California Press.

Note about the author

Junichi Murata (born in 1948) studied at the University of Tokyo, Department of History and Philosophy of Science, was a research fellow (DAAD) at the University of Cologne, Germany (1977–79). A lecturer and then Associate Professor at Tokyo University, he conducted research at the Ruhr-University of Bochum, Germany as Humboldt Stipendiat (1988–89). He is presently Professor at the University of Tokyo, Department of History and Philosophy of Science. His important publications include "Wahrnehmung und Lebenswelt" in *Japanische Beiträge zur Phänomenologie*, Alber, 1984; "Wissenschaft, Technik, Lebenswelt" in *Husserl Studies*, Vol. 4, 1987; *Perception and the Life-World* (in Japanese), University of Tokyo Press, 1995; and "Consciousness and the mind-body problem" in *Cognition, Computation, Consciousness*, ed. by M. Ito et al., Oxford University Press, 1997. His main research theme is "phenomenological theory of knowledge" in the broadest sense of the word, and has been working, on the one hand, on the philosophy of perception and philosophy of mind, and on the other hand, on the philosophy of science and technology. Both problem realms are approached from a phenomenological point of view.

Continental Philosophy Review **31**: 307–320, 1998.
© 1998 *Kluwer Academic Publishers.*

On the semantic duplicity of the first person pronoun "I"

HIROSHI KOJIMA
Niigata University, 8050, 2 no Machi, Ikarashi, Niigata, 950-21, Japan

I. Introduction

We are accustomed to using the first person pronoun "I" automatically, as it were, or almost unconsciously in daily life, convinced that we sufficiently know the meaning of it. But when we reflect upon it more attentively, it suddenly becomes clear that the pronoun "I" does not have as simple a meaning as we would think, but rather a doubled meaning that we hardly ever anticipated. The following is a rough phenomenological sketch of these meanings and of the ontological structure of the ego that reflects upon them.

On the one hand, the first person pronoun "I" means a subject that exists in the world in the plural, corresponding to the number of human bodies that exist. This first kind of I, except for a special one of them that is always called "I" by me, is usually called "he" or "she" rather than "I"; but we know *a priori* that they (he or she in the plural) are other I's too. For that reason we are never embarrassed at all, even if occasionally we hear one of them say "I" of himself or herself. This first kind of I exists not only in plural, but, as we shall see, also consciously *acts* as plural, namely is inter-subjective from the beginning. No solipsism casts its shadow on this kind of I. We would call it *"the serial I."*

I think that *das Man* of Heidegger or *die Menge* (mass) of Kierkegaard has its deeper ontological ground in this serial I. As it will become clear, neither does *das Man* exist isolated from the authentic Existence, nor does *die Menge* exist apart from *der Einzelne*.[1] Rather, both pairs (*das Man* – authentic existence; *die Menge* – *der Einzelne*) are originally and primordially included in the double meaning of the first person pronoun I.

The serial I exists as scattered in common objective space, not face-to-face, but parallel to one another. This situation also reminds us of the seriality of the practico-inert of Sartre[2] which is exemplified in the line of men on the street waiting for a bus or the invisible chain of people connected by the internet-web of computers. This is not without reason, since both the serial I and the Sartrian practico-inert have their common ontological foundation in the extensional positionality in "objective" space, which is the cardinal characteristic of the human body: the physical-body [*Körper*]. The serial I is the body consciousness proper to the physical-body, and is contrasted with the lived-body [*Leib*], the inner, more hidden, never objectifiable, side of the body.

Through the positional act in cooperation with the acts of the others the serial I constitutes the common physical-body-world. I see the world only in my perspective, but my perspective already comprises other perspectives in it. This is why I can grasp the backside of a thing that I cannot perceive here and now, and why I know *a priori* that every serial I participates in the same common world.

Such inter-subjectivity already stands before the incarnation of my serial I in my physical-body which is, for its part, constituted by this very intersubjectivity. This fact will be explained consistently only if the serial I originates from the participation of a primordial ego with primal self-identity in an impersonal, universal, positional subjectivity, which is beyond the difference of ego and alter ego. This universal impersonal subjectivity posits all physical-bodies, be they human or not, in the common space all at once and is apt to be called transcendental subjectivity, contrary to the usage of this term by Husserl.[3] And one physical-body among those many becomes "mine" only after my serial I, which always participates in transcendental subjectivity, identifies this physical-body as properly belonging to myself or as my very objectification.

Still after the identification of the physical-body serial I participates in the transcendental subjectivity which is *a priori* co-subjective. When one asks: "Where is my serial I posited in relation to this physical-body?," we will answer: "My serial I as self-consciousness is all over the surface of the physical-body." It has no view point, but a view surface. My body-surface-consciousness couples with the body-surface-consciousness of the other. This fact necessarily yields as a consequence the transcendental coupling between my physical-body and other physical-bodies around it, which is the foundation of our social life. It is the reason, e.g., why we can understand the expressions on the faces of others without comparing them anew with our expressions. The same type of bodily subjectivity already presides on both faces. This coupling becomes the necessary condition of the discovery of another serial I on and in the coupled body. Therefore impersonal co-subjectivity provides the egological inter-subjectivity incarnated in physical-bodies.

The serial I implicated in the pronoun "I" is *a priori* intersubjective. This I includes in itself another I or other I's. So it always functions as plural, though it pretends to be singular. His singularity or his difference from another is always relative; in other words, the difference is in principle perspectival under the premise of belonging to a common space. The difference can be reduced to the position of view points in the common homogenous (non-perspectival) space. When Husserl described in the *Cartesianische Meditationen* the mode of Being of another ego as "if I were there,"[4] it is clear that he was thinking it after the guiding model of the serial I scattered in homogenous space. By this method, he could therefore only constitute another I or a "second I," but could not constitute "the alien I" [*Fremdego*] for me at all, as Steinbock has justly pointed out in his book.[5]

II. The primal "I"

On the other hand, the first person pronoun "I" also means a subject who exists as only *one* in the whole world. This second kind of I has no parallel to himself in the universe. He feels himself as the sole absolute center of the universe. The universe is *his* universe: the "mine." This kind of I is expressed most clearly the moment when a person speaks to his or her rival with the words: "She or He is mine!" They, the rivals, can both understand the meaning of the word "mine," which refers to the monopoly of this kind of I. Rather, the understanding of the meaning of this word is the necessary condition of the presumable quarrel. Nevertheless, they cannot agree with one another.

On the contrary, if the serial I speaks "This physical-body is mine," reflectively intending his own physical-body, no one will contend with him. What is the difference between these two cases?

The serial I has his home only in a posited entity (*physical-body*) with limited extension, while the second kind of I, whom we would call now the *primal I*, inhabits the whole world without any restriction and limitation. He has a monadic structure, i.e., the complex of a centering body and the surrounding world, both of which interpenetrate the other. This body is totally different from the extended *physical-body* and apt to be called "*Leib*" in German, or lived-body (or even flesh) in English.[6] Just like the serial I is *a priori* the bodily consciousness of the *physical-body*, the primal I is *a priori* the bodily consciousness of the *lived-body*. But the intimacy between consciousness and the body is much stronger in the latter case than in the former case. While the serial I can be easily separated from his *physical-body* by reflection, the primal I can never reflect upon himself by distancing himself from his *lived-body*. The primal I and the *lived-body* are always one and the same.

Therefore, two primal I's can in principle never co-exist. They exclude each other *a priori* because they are inclined to occupy the absolute center of the same world and to monopolize the whole world. One might say: "This is just the expression of the original sin of Christian Theology," but we are not concerned with the theological or even the moral value of this concept. Important for us is the fact that the first person pronoun used everyday could contain in itself the meaning of such an I that is the exact opposite of the serial I.

The impossibility of reflection upon itself makes the recognition of the primal I extremely difficult, but every reflection must have by principle a non-reflective ground as its necessary starting point. If it did not have an invisible anchor in the lived-body through its physical-body, even the reflecting serial I alone could not identify its own physical-body among many co-existing physical-bodies in equality and in a parallel condition. The primal I or the lived-body is the source of my identity (I am I) and my individuality (I am not non-I).

Observed in its purity, the lived-body has no definite spatial extension or figurative shape because it is never posited in extended space. Rather, it spreads a kind of proper space indefinitely around itself (monad),[7] which is the necessary constituent of its surrounding world [*Umwelt*]. Heidegger must have been aware of the fundamental spatiality of the lived-body or the primal I trespassing the figurative limitation of the physical-body when he presented the notion of "Dasein," but at the same time, he failed to grasp the absolute identity or individuality of the primal I. This is how the ambiguity originates concerning his *Dasein* and its agent.[8]

The primal I lives only in its own monad as its unique center (mono-pole). It is no longer physical-bodies that occur within the monad, but images with proper continuity (not extension), because its monad is the proper horizon for the imagination. When 1 imagine a phantasy, my conscious horizon is closed and no one but I could enter this horizon. The primal ego is eminently the subject of imagination, while the serial I is eminently that of the perception of thetic phenomena.

The lived-body's space (monad) is not empty as is homogenous space, but rather is filled with potential Being, the Being of image. In the monad, time never flows as it does for the serial I in homogenous space, but accumulates itself into the protracted abiding present which is the proper temporality of the imagination. Image always re-*presents* [*ver-gegenwärtigt*] itself.

III. The kinaesthetic unity

We have brought to light two different meanings of the first person pronoun "I." They seem to be extremely different from each other, but in reality, these two meanings together constitute a unity called "I." How is this possible?

Incidentally, the Japanese philosopher, Nishida, insisted upon "the absolutely contradictory self-identity" of the human being. Perhaps after having discovered the duplicity of the meaning of ego as described above, we could no longer take his expression as simply strange or extraordinary.[9]

What is the inner relation between these two I's, the serial I and the primal I? The primal I in its primordiality accepts the transcendental subjectivity that is impersonal and then suffers its *Verköperung* or objectification of its lived-body in order to yield the serial I (physical-body = ego) as its self-reflection. This reflection begins generatively from one's birth and is genetically reaffirmed every morning upon waking from sleep. For sleep is considered to be the periodical retreat into the pure primal I.

Then this self-reflection of the primal ego is promoted and completed by the serial I through the identification and assumption of a certain physical-body precisely as its "shell," and through the constitution of body-surface-consciousness (formal incarnation). This physical-body of the serial I, however, is fulfilled at the same time by its lived-body or the primal ego from the "inside" once again (material incarnation).

In this way, in the process of the self-objectification by an alien subjectivity and the re-incarnation by itself, the primal I reaches unity with itself, but only with itself as alienated and objectified. This is the complex unity of the serial I and the primal I; that of the physical-body and the lived-body, which is to be called the *kinaesthetic unity* of the ego.

It is characteristic of kinaesthetic unity, which is the most ordinary figure of the human ego in co-existence, that the serial I is still objectifying the primal I, keeping the lived-body under the restriction of the physical-body. The lived-body is never objectifiable in itself, but now spatially molded into the positional extension of the physical-body, and cannot retain its primordial unity with the surrounding world. The lived-body is virtually posited Here in the common empty space (thrownness). However, this Here is not the absolute Here that the monad of lived-body originally occupies, but is only the relative Here that can be indefinitely pluralized as Here = Theres.

Many philosophers including Husserl expressed this plural-polarization of Here (lived-body) as "empathy" or "introjection," but if this word means a willful throwing of my absolute lived-body into other physical-bodies, it is incorrect. Such an expanding penetration of my absolute lived-body never occurs, not only in the case of empathy (of another ego), but also not in the case of sympathy (e.g., of sorrow, etc.) because sympathy is necessarily preconditioned by the empathic discovery of another ego. In any case, since imagined Being has only a possible Being and never reaches the real Being of others, I can never imagine the real existence of another kinaesthetic I by myself.

Rather, once my primal I accepts his *Verkörperung* or objectification by transcendental subjectivity and, reflectively objectifying itself, is unified with itself as the kinaesthetic I, then my monadic lived-body is passively and automatically polarized at once into plural lived-bodies presiding in coupled physical-bodies around my physical-body. This is not an active imagining, but a kind of passive synthesis of plural reality.[10]

Alfred Schutz was right when he admitted the existence of the other stream of consciousness in a form similar to mine behind the moving physical-body in his *Der sinnhafte Aufbau der sozialen Welt*, but it is remarkable that he abandoned any attempt to give a foundation to it.[11] The existence of an immanence other than mine will be founded only by the passive pluralization of my lived-body through the alienating *Verköperung*, or objectification of it.

Thus, the kinaesthetic I as lived-body = physical-body is an incomplete unity of the serial I and the primal I. It cannot mediate the primal I in its monadic totality with the serial I. Rather in the kinaesthetic dimension the serial I and his positional principle predominates over the primal I and his solipsistic principle. It is clear that language also has its ontological foundation in this kinaesthetic unity of I in which the intersubjective principle predominates. It therefore seems probable that the primal I must supplement the linguistic expression of his "only-oneness in the world" (monadicity) with the phenomenon of mood. For mood is the appresentative appearance of the concealed lived-body = monad to the kinaesthetic I, which appearance always shades off according to the accessibility to its monad in this situation (e.g., from complacency to boredom, melancholy, anxiety, etc.). In fact, many people including existentialists and Jungians think that the first person pronoun "I" could not adequately mediate the individuality of human being, but needs another term to express it properly (e.g., the self or existence).

So it will be possible to doubt whether even the phrase, "He/She is mine!," which was referred to as expressing the monopoly of the ego, has meaning without the understanding of a specific mood accompanying it.

Nevertheless, we could not deny that the pronoun "I" sometimes implies in itself the individual aspect of personality as its meaning, especially contrasting this aspect with the intersubjective aspect of it. For example, if anyone seriously says: "I love him or her," this "I" never means the serial I as one among many human subjects, but means precisely the one and only I in the world: the primal I.

So we must ascertain here more clearly whether or not the kinaesthetic unity of the I could in some way contain the totality of the primal I in itself.

IV. The open image

As the kinaesthetic I is the proper inhabitant of the lifeworld, we must initially address our attention to the structure of the lifeworld. Just as the kinaesthetic I has double aspects – physical-body and lived-body – so too does the lifeworld have double aspects – extensional perspectival space and the open field of image.

When viewed from the side of physical-body, the lived-body within the lifeworld is divided and discontinuously molded into each physical-body with limited extensionality, but once viewed from the inside of the lived-body, the lived-body as a whole constitutes a continual field (terrain) underlying extensional space and time. This terrain of the lived-body is a pre-given and pre-thetic moment of the lifeworld, and every image in it belongs to the field itself and not to any particular subject.[12] Unlike free phantasy in a monad, these open images have a strict correspondence with spatial extentionality, For example, when I feel a pain, it is originally the quality pain itself that belongs to no one in particular. Only once it is located in my physical-body does it become my pain (or more exactly, a pain originated in me). The same pain can be a pain that originated in another person when it is located in his or her physical-body through various external symptoms. But in both cases I feel the same quality pain, only in different degrees and perspectives.[13]

We can say the same thing about the open images of color, tone, smelling, taste, and touch. For each kinaesthetic I, they have a common neutrality and, once connected with spatial extentionality, they are called 'sensuality.'

We admit that there is a difference of degree in these images according to the individuality of the person, but a kind of normality is always presupposed for such images and accordingly for sensuality which allows for individual diversity. But beyond the range of normality, the difference of degree in sensuality is regarded as abnormal.

We have now another, third, type of image that should be located between the open image in the neutral field and free phantasy in the closed monad. For example, the image of sorrow could indeed be indirectly connected to the expression of the face, but could not be directly connected to any spatial extension; rather, it is directly connected to other open images, e.g., that of pain. We can say the same thing about the images of pleasure, anger, likes, and dislikes, love and hate. Such an image, like sorrow – which is the double image with respect to the open image – is generally called "emotion." There is much more diversity between individuals where emotion is concerned than, say, sensation, such that to possess a common emotion (sympathy) would mean to live in a restricted common field which is the special lifeworld. This special lifeworld is not as open as that of sensuality, and not as closed as the monad.

The emotional image comprises in itself a so-called value judgment, while the sensual image comprises only the judgment of fact. Therefore, the special lifeworld founded upon the sympathy of a common emotional image is usually guided by a certain common value judgment. We could see many good examples of such community in the Modern National States. Race is originally a complex of sensual images (e.g., the color of skin and hair, the manner of speaking, the beauty of women, etc.); nationalism is based upon this when emotional images are added. The emotional image has, as we have seen, originally no proper correspondence in the external world, but, when it is combined with sensual images, it often produces an artificial correspondence in the outer extensional world which is called "symbol." The special lifeworld often has its own symbol that is regarded as sacred for its members.

The sympathy caused by some common emotional image is sometimes accompanied by the antipathy against different emotional images cherished in the respective sensual image. Then the special lifeworld (community) supported by sympathy becomes more confined and more aggressive to the other special lifeworlds, as many examples have shown (e.g., in the case of the Jewish-Arabic conflict).

As already shown above, mood is the appresentative reflection of the hidden monad of the primal ego upon the open image-field of the kinaesthetic ego. This appresentation, however, is never without nuances, but is always colored by the accessibility to the monad, which from time to time is thrown or pro-jected into the future. While I grasp my own mood directly through the whole horizon of my living world, the mood of others is given only through the bias of the coupled kinaesthetic bodies.

Therefore, the open image-field as a place of reflex of the monad is colored by my mood, which is however also under the influence of another's mood. When I feel melancholic, my projection is threatened to fail, the common lifeworld itself also becomes somewhat blue. But, on the other hand, the open image-field for its part can influence the mood of an individual. For example, the look of faces of other people, the color of the wall or the sound of background music could to some extent change the mood of individuals and could contribute in making a certain common mood that is usually called "atmosphere." Of course, it is also possible to reject the atmosphere around me and to keep my own mood and to confine myself to it.

V. Projection and meaning

The subject of projection is the primal ego and not the kinaesthetic ego, but as already suggested, the on-going projection of the primal ego reflects upon the open image-field of the lifeworld. The accomplished project reflects upon it as a mood of satisfaction, while the disrupted projection causes a mood of melancholy.

But the influence of the primal ego upon the kinaesthetic dimension is not limited to the genesis of mood. The concealed projection of the primal ego gives to the open images of the lifeworld a kind of rearrangement, discriminating between them what is suitable to his projection and what is not. Then the former is necessarily accompanied by the emotional image of likes or pleasures and the latter is accompanied by that of dislikes or hates. So, a considerable amount of emotional images, (though not all of them), are caused in the open lifeworld by the concealed projection of the primal ego. It is also clear that this projection is the source of the value judgment mentioned above. Accordingly, sympathy and antipathy, which promote the constitution of the special lifeworld (community), are often undergirded by a latent projection that is shared by members of that community. But the freedom of decision concerning this projection is rarely secured for the individual (the primal ego) in such a partial community.

Up to now we have inspected some aspects of the correlation between the kinaesthetic dimension and the primal ego. But the most important one remains untouched: the relation of the linguistic to the semantic.

The total Being of the accomplished aim of project [*Worumwillen*] could itself in some way leave a trace upon an open image as its meaning because every image in the open field could be transferred into the monad,[14] and once the total Being of the monad is converged into this image, become the aim of projection. Due to the convergence of the total Being into an image there is necessarily a pro-jection of that Being into the future. In other words, contrary to a Heideggerian interpretation, every open image could be a *Worumwillen* of a projection once transferred into the monadic dimension.[15]

In fact, not only the open images, but also the kinaesthetic successive protentions are also able to be absorbed into the primal ego who makes this protention successively accompany, as well as adjust to, the tool-nexus of his projection of the monad; but, on the other hand, it is not easy to absorb the total Being of a monad into the open field of the kinaesthetic terrain since the detotalizing force of transcendental subjectivity already dominates this area. Here a certain mediator (to be discussed later) would be necessary, a mediator who could keep the totality of the monad from any detotalization.

This trace of projection in the image is not a focus of phenomenal unity of that image, but a focus of semantic unity of it, namely, what it is for. It remains as it is, even when the open image in neutrality is put together synthetically into the optimal image for every one. Then it is aptly called the noema of this image. Namely, noema is the semantic focus of the optimally generalized image for every kinaesthetic ego. In fact, the open image of pain as quality is neutral and has no peculiar possessor, but it might have some difference of degree in it according to the difference of perspective. The

[91]

optimal image of pain is a typified pain trespassing the difference of degree and perspective in it. It is the de-perspectivation and omni-localization of image which is made possible only by the concentrating synthetic force of transcendental subjectivity that dominates the terrain of open images through the kinaesthetic ego.

The noema is not yet a linguistic meaning, but rather its pre-stage, omni-localized semantic focus of the non-perspectival image. In order to establish linguistic meaning, such omni-locality of the lived-body-element of the kinaesthetic ego would have to be combined with the omni-temporality of the physical-body-element of the kinaesthetic ego which is concretized as kinaesthetic schema in the vertical intentionality of that ego (the non-dated memory). Not only are they combined as a mere complex of the omni-temporal physical-body and omni-local lived-body, but the physical-body schema would also have to become omni-local and the noema would also have to become omni-temporal in a synthetic combination. Then the linguistic schema and the linguistic meaning are born, both in their omni-temporality and omni-locality, where the former is necessarily the phoneme or grapheme which is the metamorphosis of the spatial extentionality proper to the open image.[16]

The promoting power of this omni-temporalization and omni-localization toward the linguistic dimension might seem to be the impersonal transcendental subjectivity in general, which is also omni-temporal and omni-local. But the noema as the origin of the linguistic meaning has its ultimate source in the aim [*Worumwillen*] of projection of the primal ego pertaining to the open image who, as ego, has as such no direct relation with transcendental subjectivity. We are therefore required to find a mediator who could mediate between transcendental subjectivity in general and the primal ego in totality, because any projection demands the total (monad) of the primal ego. The kinaesthetic dimension, which is from the beginning under the control of transcendental subjectivity, is evidently not able to mediate between them. It seems to us that here lies the deep mystery of the human language which is beyond the scope of transcendental analysis.

We referred to the inability of language to express the individuality of the primal ego which might be seemingly supplemented by mood. But already whenever we say in a poem: "a house" or "a tree," does not something more than the mood of my primal ego occur here between the I and images? Whenever the meaning of these words comes to us, is there not a change within us which, allowing for my individuality, nevertheless mollifies the confinement of my monad?

If we are correct in this line of thinking, the duplicity of the meaning of the first person pronoun "I" is accomplished not in the kinaesthetic I as such, but only in the kinaesthetic I mediated by the still unknown mediator. Only in the latter case, the mediated kinaesthetic, could I express adequately either the serial I who is as numerous as there are bodies, and the primal I, who exists as the only one in the world.

For this mediated ego, to give a name to a thing would mean not only to objectify it or to grasp it in a universal concept, but also to recognize it as a monad, as an individual totality that is proper to it.[17] On the contrary, for the kinaesthetic ego as such all the thing seems to lose its individuality, and becomes more and more a mere aggregate of phenomena without Being, as it is ordinarily seen in scientific investigations.

VI. Conclusion

The serial I is the self-consciousness of the physical-body who exists in the plural, not in a face-to-face relation, but only as parallel to another. They are first of all he or she for me, and I am also he or she for them. Therefore, the serial I is the aspect of the first person pronoun representing the third person dimension.

On the contrary, the primal I is the genuine first person dimension without whom even the serial I could not stand, as was already shown. But this ego has no partner at all; for he is absolutely alone.

The kinaesthetic I is the incomplete unity of both egos mentioned above, ego who could indeed take or exchange each other's role through the associative coupling of bodies in the coexisting lifeworld, but it knows no encounter in totality.

The problem concerns the mediation of the first person dimension (the primal ego) and the third person dimension (the serial ego), how one can be mediated with the other completely, either inside or outside of the primal ego, as the genesis of language will require. For the serial I is inside of the primal ego insofar as the former is the constituent of my kinaesthetic ego, while the serial I is also outside of the primal ego insofar as the former is the constituent of the other kinaesthetic ego. The double mediation in both – inside and outside – by the same mediator will be the necessary condition for the birth of speech, which, once generated inside of me, should be transmitted to outside of me, or first transmitted from the outside, should be apprehended inside of me.

It seems quite natural that in this case we pay attention now to the *second* person dimension that has not been treated at all up to this point.

Why does the second person dimension exist? Do not the first and the third person dimensions suffice? Is it necessary for us to use the word "you" to designate the other ego? Are the words "Mr. A" or "Mary" not always substitutable with "you"?

[93]

Considering these questions, we cannot help referring to Martin Buber's definition of the second person dimension. He says: "The basic word [*Grundwort*] I-Thou can be spoken only with one's whole Being. The basic word I-It can never be spoken with one's the whole Being."[18] We think that what Buber wants to ascertain is that only in the second person dimension is to be found the total, genuine encounter, and never in the third person dimension. But we must add: nor in the first person dimension, since (as already shown) the first person dimension is a monad or a kinaesthetic agent that has no ability to produce a genuine encounter. For this reason, referring to Buber, the word "you" must implicate something particular that the word "Mr. A" or "I" (even "another I") could not mean.

But, as mentioned above, the total encounter we need is not only the encounter between my ego and an outside alter ego, but is also the encounter between my (primal) ego and an alter (serial) ego inside of me. In this respect, Buber's interpretation of the I-Thou relation is very inadequate because he regards it only as a one-way relation: as a relation that only starts from the I in totality. But the genuine encounter should be a reciprocal relation, starting from both sides at the same moment. Thus, the I-Thou relation in this sense must at the same time be the Thou-I relation. In other words, the You in front of me is an alter ego at the same moment, and I am also You for this alter ego. Every I-Thou relation is in itself reciprocally duplicated:

I (Thou) —— Thou (I = alter ego).

Should anyone think that the I and the Thou occur successively in turn, he or she would not understand what the encounter is. Rather, the complete coincidence of I and Thou on both sides is the necessary condition of the genuine encounter. So we have a new diagram:

Thou = I —— I = Thou.

This diagram shows that through both my kinaesthetic ego and alter kinaesthetic ego, the same Thou occurs as follows:

Thou >> I —— I << Thou.

and meet Thyself again between them:

I >>>>> Thou <<<<< I.

According to our interpretation, this is the structure of the genuine, total encounter between my ego and alter ego. Thou appears not only in front of me as my partner, but also appears in and through myself as the kinaesthetic ego and mediates my total primal ego and my serial ego inside of me, while by the phenomenal self-encounter between I and another I, Thou mediates both Is, where another I (alter ego) is always able to be called He/She, too.

Such a total encounter occurs not only between human beings with speech, but also between an artist and natural things, where they appeal to the former through the occurring Thou to give them eternal figures, and the artist responds to it with creative activities promoted by Thou from the inside.

We could therefore draw the conclusion that the unknown mediator who would mediate the first person dimension and the third person dimension completely is the second person dimension, *Thou*, who is now grasped in a much wider perspective than before in the case of Buber.[19] Thou is not only the partner of the I, but also the sole mediator of the inner duplicity of every ego. Even the totality of Buber's I who should speak the basic word I-Thou is already mediated by Thou.

The kinaesthetically uttered first person pronoun "I" could adequately imply in itself its primal I as a monadic totality (the source of meaning), only when he (the kinaesthetic I) is mediated by the second person dimension, "Thou," or (what is the same thing) is totally encountered by another being. Then and only then is accomplished Nishida's "the absolutely contradictory self-identity" of the human being.

Notes

1. Cf. Martin Heidegger, *Sein und Zeit* (Tübingen 1953), 126 ff., and Soren Kierkegaard, "Der Einzelne," in *Die Schriften über sich selbst* (Gütersloh 1985), 96 ff.
2. Jean-Paul Sartre, *Critique de la raison dialectique* (Paris 1990), 165 ff.
3. Husserl always confined transcendental subjectivity to the human ego. Cf. Edmund Husserl, *Ideen I*, Husserliana III (The Hague: Nijhoff, 1950), 300.
4. Cf. Edmund Husserl, *Cartesianische Meditationen*, Husserliana I (The Hague: Nijhoff, 1963), 147.
5. Anthony J. Steinbock, *Home and Beyond: Generative Phenomenology after Husserl* (Evanston: Northwestern University Press, 1995), 57 f.
6. Here this term is used only for its pure mode without any restriction by the physical-body.
7. For a detailed analysis of the monad as *Leib*-space, see my book, *Monad and Thou* (in preparation).
8. Heidegger's *Dasein* is implicated in either *das Man* within a mass, or authentic existence as *solus ipse*.
9. Cf. Nishida Kitaro, *Zenshu*, vol. 9 (Tokyo, 1949), Chapter 3.
10. This kind of passive synthesis seems to be difficult to approach by the transcendental-phenomenological method alone. Cf. Natalie Depraz, *Transcendence et incarnation* (Paris, 1995), 125 ff.
11. Cf. Schütz, *Der sinnhafte Aufbau der sozialen Welt* (Frankfurt a.M. 1974), 137 f.
12. It could be identified with the pre-given world to which the later Husserl referred in many places, though he seems never to have reached a definitive conclusion about its relation to imagination.
13. Therefore, the idea that we can feel the pain of others only through introjection is erroneous. Cf. Max Scheler, *Wesen und Formen der Sympathie* (Bonn, 1985), 238 ff.

14. The fact that an image happens to belong simultaneously either to an open image field and to a closed monad is the ultimate source of laughter.

15. For Heidegger, the *Worumwillen* of a projection is my Self discriminated from images of tools as the *Woraufhin* of the projection. Cf. Martin Heidegger, *Sein und Zeit*, (Tübingen: Niemeyer), 86.

16. It is characteristic of the open image that an extensionality strictly corresponds to it.

17. Cf. R.M. Rilke, *Duineser Elegien* (Wiesbaden, 1955), Die neunte Elegie, p. 35.

18. Martin Buber, *Ich und Du* (Heidelberg, 1979), 9.

19. Regarding the detail of my concept of Thou, see my book, *Monad and Thou* (in preparation). It is remarkable that my concept of Thou is very close to that of Buber's *Geist* (Spirit).

Note about the author

Hiroshi Kojima, Professor Emeritus of Niigota University, (born 1925 in Tokyo) has taught Philosophical Anthropology in Niigata University from 1982–1991, and has received his title of "Doctor" (LitD) from Kyushu University in 1992. With his main interests in phenomenology, and existential and dialogical philosophy, Kojima has translated Martin Buber's *What is Man?* (1961) and *Encounter* (1966) from German into Japanese. One of the founders of the Phenomenological Association of Japan (1980), he has remained an inspiring presence in conferences featuring an "East – West" theme, has edited *Phänomenologie der Praxis im Dialog zwischen Japan und dem Western* (1989), has co-edited *Japanese and Western Phenomenology* (1993), and has contributed articles to *Analecta Husserliana*, to *Philosophisches Jahrbuch*, as well as the to the *Encyclopedia of Phenomenology*. He is currently preparing a book for publication in English, entitled *Monad and Thou*.

Continental Philosophy Review **31**: 321–335, 1998.
© 1998 *Kluwer Academic Publishers.*

Qi and phenomenology of wind

TADASHI OGAWA
Graduate School of Human and Environmental Studies, Kyoto University, Kyoto 606-01, Japan

Introduction

A famous Zen priest Hohtetsu was gently fanning himself. A young monk asks him: The wind-nature is always everywhere between heaven and earth. Why are you using the fan? The priest answers: You know only that the wind-nature is always everywhere. However you do not know the way-truth that no place is not penetrated by the wind-nature. The monk asks: What is this way-truth? Then the priest answers without a word using only his fan. Dogen explains this dialogue: To say that you do not use the fan because the wind-nature is always everywhere, and that you shall hear wind by using no fan, means that this young monk does not know the always and everywhere being of the wind-nature. The wind-nature is always everywhere, therefore the Buddhistic wind lets the golden earth be. . . .[1]

What is the wind-nature in this story of the famous zen-buddhist priest in Dogen's *Shobo-Genzo*? My answer is that it is qi/ki. The word "qi" is originally from China. This word is now perfectly Japanized. We Japanese make use of this word in everyday colloquial conversation. But we are normally not conscious of the original meaning of the word: My friend is ill ("*byoki*"); his speech has no vitality ("*seiki*"); his eagerness ("*shiki*") has faded; his acts seem not to be vigorous ("*seiki*"); he is now "*inki*," his "*yo-qi*" (*yan-qi*) is gone. The word "ki" in these sentences is the same as the word "qi." I can also make these Japanese compounds: "*kyo-ki*" (madness), "*kekki*" (hot blood), "*doki*" (anger), "*sakki*" (sanguine). These Japanese words are variations of the same word "ki." An arrangement of these words shows a risky atmosphere.

I will elucidate the broad meaning of this word "qi/ki" in the following discussion. The range of senses of the word qi is actually vast. It is not easy to grasp the essence of qi, for it is ambiguous. You will see that qi "fills in" both an individual body and all that is between heaven and earth.

In the latter sense, the qi of the world in our times is not in order but is disturbed. World politics has been upset since 1989. The atmosphere and mood of the world is always tumultuous. Due to this disorder of world-qi, earthquakes, climate disorder, and civil wars in the Balkans and Africa seem to have occurred.

But how can one say that the qi between heaven and earth and the qi in the human body are the same? Is there not some great difference between these aspects of the appearance of qi?

I. The meaning of qi

What is the meaning of qi? The Chinese word qi/ki means something natural around the human being, between heaven and earth like vapor and steam, or mist and fog. Qi is air, weather, climate, atmosphere, and so on. Qi/ki is a kind of steam surrounding human beings, and has the special meaning "mood" when this qi/ki appears in the surroundings of the person.

A Japanese and Chinese expression like *fuhtei* (literally: wind-body) denotes this mood of a person surrounding him or her. Therefore qi is a kind of wind, the wind of breathing. This type of qi is ex-piration and in-spiration. The air of exhalation and inhalation, the wind of a person, which means the wind-body, is really the way of living. Qi is the air and the wind. The inspiration and expiration, the wind-body of the person, is nothing but the movement of which the human being is directly aware in his or her body. This awareness is grasped intuitively and directly like when we say in English that "the atmosphere is tense" in the sense that "trouble's brewing!" or like one says in German, *"dicke Luft."* This atmosphere is expressed in Japanese as "a tense air" [*kinchosita kuhki*].

Mencius first realized the deep and relevant meaning of qi for the world and human beings. He said, qi "fills in" the body and is commanded by the will. Mencius, the voluntarist, was inspired by Daoism (Song and Yin) and expressed this sense *"hao-jan chih ch'i"* (*"kozen-no-ki"*): the strong moving, spiritual power of the human being.

"Kozen-no-ki," according to Mencius, is difficult to express. This qi/ki is greater and larger, wider and stronger than anything. If this qi/ki is spread out in a linear fashion it fills up the world between heaven and earth unto the very end of the world. There are two things that occasion this qi/ki. The first is the "rightness" as an accumulation of just deeds. The other is the empty and formless way that unfolds between heaven and earth. This is the truth of the way which is the opening up of the world. Qi/ki therefore has two major meanings: On the one hand, it is the living and moving power, the vital, spiritual power that is embodied in the human body; and on the other hand, qi/ki is the power between heaven and earth, the vital and spiritual air. But how can the two meanings of qi/ki be combined with each other?

According to Mencius, the answer is the will. He thinks the will is the unity of the two meanings of qi/ki. But as Shigeki Kaizuka, a famous Japanese Sinologist rightly says, the function of this will is not elucidated adequately by him.

Since antiquity it has been said that qi/ki is not in order, but is disturbed. Today we speak this way about the dimension of the personal body. But our ancestors used to use this expression not only to speak about disturbances of the human body, but also to characterize a prompt change of the qi/ki in the world.

The concept of qi/ki in Mencius played an eminent role in Japan before the Meiji-Restoration. For example his successor, a patriot poet in Nam-Song, Wen T'ien-Hsiang (1236–1283) composed the poem "Cheng-ch'i-ko," the song of the right qi/ki. Responding to this patriotic poem, our thinker-poet, Toko Fujita, who was a prominent politician in Mito, wrote the poem, "A response to Wen T'ien-Hsiangs poem "Cheng-ch'i-ko"." This poem of Toko Fujita inspired many royalists of the Restoration period. Among these noble minded patriots Genzui Kusaka of Nagato, Choshu was a very famous disciple of Shoin Yoshida, one of the famous theorists of the Meiji Restoration. This poem of Toko Fujita inspired these noble minded patriots.

As the late famous novelist, Ryotaro Shiba has emphasized, it was not in vain that the thought of the post-Mito-School arose, and a poet-thinker like Toko Fujita was able to place our country in a spiritual patriotic situation. The important position of Japan today depends on the fact that Japan was not invaded by the Euro-American peoples, and that we unified our national State quickly, though it can be said that the speed of this establishment of the State also caused some problems. What made this possible? I contend that it was the political atmosphere before the Meiji-Restoration, the ideas of which came from the Post-Mito-School. However, the purpose of my paper consists not in describing the political sense of this School, but in interpreting the sense of the world from the perspective of qi/ki.

According to the poem of Toko Fujita, the Japanese upright or true qi/ki appears as the Japanese spirit, as the Japanese sea, as cherry blossoms, as Mt. Fuji and as the Japanese sword. Qi/ki in this sense is the original power that changes its form according to situations. Qi/ki is between heaven and earth and is living in the human body, animals, and plants.

In the Mito-School, the qi/ki of Mencius and Wen was quite Japanized. Qi/ki is used as an ontological concept, and it means human interrelations. A human being is born by receiving the body of his father and forefathers and by receiving the qi/ki (atmosphere) of heaven and earth. "Heaven and earth, the father and the forefathers, are the origin of a man" (Aizawa).[2] Father and son arise from the same origin, namely, the same qi/ki. The same river makes many tributary streams and waterways flow into the same river. Father and son are different in body but the same in qi/ki. Our bodies are the remains of our father and mother. In qi/ki they are the same. Aizawa put it this way: "*Fu-shi-i-kki*" ("The father and child are in the one and same qi/ki"). If you want to see your deceased mother or father you can see her or him in a mirror. The faces of one's parents appear in the mirror. Aizawa cited an old song in the Heian period: "A child of a man would like to see him in the mirror when his heart aches for his dear parents."[3]

Father and son are different in body but unified in qi/ki. A family is in the same qi/ki, a nation has the same qi/ki, and therefore the same unified qi/ki-body. The supposition that a nation has a body like a human being allows us to speak of the body of a nation (Yasushi Aizawa). This analogue between a human being and a nation is based ultimately on the national theory of qi/ki-metaphysics. (I would like to note as an example of intercultural invariance[4] that Plato and Aristotle, even Hobbes, quite often employed this analogy of the body-*polis* [city-state]).

II. Intercultural phenomenon of qi/ki

You may object that this concept of qi/ki is a product of typically East-Asiatic thinking. I will ask, however, if only East-Asiatic thought conceives of qi/ki. Is it true that Europeans and the Near-Eastern peoples do not know about this phenomenon? To this I would respond that all humanity is aware of this phenomenon. Qi/ki is the phenomenon of breath, expiration and inspiration; the Greeks and the saints of the New Testament called it *pneuma*. *Pneuma* is the wind. Jesus Christ said to Nicodemus: "The wind blows where it will, and you hear the voice of it, but you do not know whence it comes or whither it goes; so it is with everyone who is born of the Spirit" [*to pneuma hopou thelei pnei, kai then phonen autou akoueis, all' ouk oidas pothen erchetai kai pou hypagei. houtōs estin pas ho gegennemenos ek tou pneumatos.* John 3:8].

The last words "*ek tou pneumatos*" in this American authorized translation is mistranslated. (And I have to change the word *fone* to "voice" rather than this American authorized translation of "sound"). Wind and spirit are originally the same word: *pneuma*. The translator translates the same word, *pneuma*, with different words arbitrarily depending on the context. To express the natural phenomenon of "wind" there is another word, "*anemos.*" In reality Jesus used the same word *pneuma*, seeing wind and spirit as the same. *Pneuma* is the breath of God who appears as wind and as spirit.

I understand these words of Jesus to mean that persons are human beings like the wind, they are the children of the spiritual wind from God. We persons are spiritual human beings like the wind. Persons draw breath, breathe in the world between heaven and earth.

Prior to this *Pneuma*-thought of Jesus, a Greek *pneuma*-thinking was prepared in the Mediterranean world. It is through breath that Plato analyzes the relation between the fire in the body and the cold environmental world. For example Plato elucidated that the human being is the existence of "qi and wind" in the high level from the viewpoint of breath and qi. Through in-spiration and ex-piration the body acts and receives air in a passive way. What is working in the body actively and passively makes the body cooler and wetter with the cool air from the world. Breathing in this way, the body is able to get nourishment and to preserve life since, according to Plato, the fire in the body moves with the air of breath, digests food and assimilates it to itself. The movement of qi, namely, breath, mediates between the cold world and the glowing fire in the body and it keeps the life-fire burning in the body.[5] Plato explains how the human living body is rooted in the cool world for itself. The life-fire in the living body belongs to the world through the breathing of qi. It is precisely qi that unifies the body with the world.

You may object to this explanation of "qi in the body and in the world," thinking it to be an old metaphysical idea concerning the relation between microcosmos and macrocosmos. But this objection is not correct. On the contrary, it is appropriate to accept the idea that qi appears as air, wind, and breathing. Qi is given with the body of the human being. Qi is opened up from the human body. Hence, Plato understood breathing as the relation of the body and qi, expressed by the Greek words *"to ergon/pathos,"* signifying the active and passive modes in the body. The human being is a being like the wind. I would now like to develop a phenomenology of wind.

Wind does not only play a beneficial role, since the wind, especially the cold wind, can do harm. Qi/ki has two aspects, *yan-qi* (*yo-ki*: cheerful atmosphere) and *yin-qi* (*in-ki*: gloominess). The wind blows merrily and gloomily. The cold is the result of the cold wind. Therefore Baudelaire sings in his poem: *"Hélas! tout est abîme, – action, désir, rêve,/Parole! et sur mon poil qui tout droit se relève/Maînte fois de la Peur je sens passer le vent."*[6] Wind is really "the wind of transitoriness" according to Buddhist saint Ippen.

In Japan people believed that the wind blows at the crossroads because evil spirits were there. You meet something strange and forbidding in the intersection of two ways. The Buddhist statue Jizo-bosatu was set up against the wild and evil wind.

III. A possibility of phenomenology of wind

In the phenomenological movement of today, much emphasis is put on thematizing the dimension between person and things, between subject and object. Concepts such as *subject or object no longer play a relevant role.* "Passivity"[7] in Husserl, mood [*Stimmung*][8] in Heidegger, and "Atmosphere"[9] in Schmitz, and last but not least "Fundamental Mood" [*Grundstimmung*][10] (Klaus Held) are key concepts in phenomenological thinking. According to this analysis of phenomenological thinking, concepts like subject and object are too broad and too influenced by dualistic-substantialist thought. Because of this substantialism, the human being is seen as a subject isolated from the world, and the thing is seen as an object autonomous and independent of the perceiving subject. There is a world between human being and thing. This world, this dimension, is filled by something that is neither human being nor thing. What is this something? This is properly speaking "mood" or "atmosphere." This type of thinking is originally found in East Asia, too – qi/ki.

Heidegger's "mood" and Schmitz's "atmosphere" are discovered in the background of the opening-up of the coexistence or co-penetration of world and I. Mood and atmosphere make it possible for human beings to be in the world. These phenomena preceding all other moments of the lifeworld let them open up precisely the world. As Husserl investigates the pre-predicative logic of the world in this fundamental dimension in which the world and the I are interfused with each other, this dimension has the role of discovering the truth and logic of the world. Everyone seized by a mood or by an atmosphere understands and thinks. For this reason Heidegger writes understanding is always "tuned" by a mood.[11] Thrown into a mood and an atmosphere, everyone – finding himself or herself in a situational mood or atmosphere will be projected to understand and to elucidate the sense of being.

Based on his reinterpretation of Heidegger's concepts, "*Stimmung*" and "*Grundstimmung*," Klaus Held, first among todays phenomenologists, recovered the relevance of mood for the disclosure of truth: evidence. The appearance of the world relies on mood; the world opens itself up through mood. "Existentially, a temperament implies a disclosive submission to the world, out of which we can encounter something that matters to us."[12]

In the following consideration I take the technical term mood [*Stimmung*] in the sense of atmosphere. To my mind, mood and atmosphere are the same. And I think that qi/ki is to be conceived along the same lines.

The meaning of "qi" has connotations of "energy" and "power," implying naturalistic and substantial elements. If we are able to purify "qi" of this naturalistic connotation, giving it the sense of appearance, it will only be thanks to the phenomenological epoché. After this phenomenological method we will be able to change "being" into something appearing, and then we will be able to elucidate mood and atmosphere in a genuinely phenomenological manner.

What is "atmosphere"? Atmosphere and qi/ki are the same in the sense that both mean the sphere of mist or steam around a person. The atmosphere or qi/ki has no identifiable point or place in space. It overflows and spreads out over the world as a whole to the very end of the world. Atmosphere grasps the human being at the root of his being. The root of his or her being is the body as a limited place of atmosphere. A person has his or her body-feeling in the awareness of his or her body-states: hunger, fullness, thirst, fatigue, freshness, e.g. the living qi/ki ("*seiki*") in the body. A person is always seized by atmosphere through his body-states and is swallowed by atmosphere. This idea is found in the phenomenology of climate in Schmitz. If you have ever experienced the summer in Kyoto, you have only to remember the oppressive weather at the end of rainy season just before the height of summer. This heavy weather comes from heat mixed with humidity. This oppressive weather, together with our body-states (gloominess), crushes everyone's spontaneous will.

You may object that this thought is irrational. The spontaneous will in the human being is so important that it cannot be ignored. Anyone wanting to think rationally will say that this is treason to rationality! But if you focus on the things themselves you will see that in relation to atmosphere, which can seize hold of us and swallow us up, the human being is actually powerless. The idea that the atmosphere controls human beings is as old as the Greek world in the Homeric era. Or, think of the situation at a football or tennis match. The players and the audience are melted into one and the same situation. Each person is embedded in the enthusiastic situation of every fine play. This situation is evidence that atmosphere grasps every human being.

The famous French sociologist, Emile Durkheim, elucidated the following thesis in his study, *Les formes élémentaires de la vie religieuse*: Society can exist in and only by individual consciousness. The collective power penetrates into individuals and is organized in individuals. Power becomes the essential ingredient of the individual being, and through this, collective power becomes stronger. This Frenchman has remembered the French Revolution. The individual is embedded in the mass, and the mass penetrates individuals. This interaction among individuals in society takes place quite often and transmits the affective effects to one another. "Individuals seek each other, make meetings more often." Affection strengthens itself. Spiritual and emotional energy covers the people, arises in oneself, echoes in him or herself and strengthens his or her activity. "It is no longer the mere individual but the mass who speaks. The mass is incarnated in the individual and is personalized in oneself."[13]

What then is atmosphere? In the totality of atmosphere, there must be a melting space of things, human bodies, and human perception. This melting space has three essential elements that I will explain under the following points: (1) the radiation of things, (2) synaesthesia, (3) the feeling of body-states.

Radiation of things

Atmosphere fills human beings and their environments prior to the independence of persons on objective things. The philosophical tradition presupposes the separation of the inner and outer world in human beings. We, as human beings, think that we have an inner world separated from the outer world. However we have no inner world since we are always embedded in situations. We find ourselves in our situation. As Heidegger and Schmitz have shown, mood and atmosphere are not subjective feelings of mental events that one could find in an inner world or perhaps in consciousness. "Mood [*Stimmung*] comes over human being" as Heidegger writes. Mood does not come from the inner or outer world of human beings, but comes-over from one's Being-in-the-world. The affection of love, the atmosphere of love, does not exist in the individual person, but appears in the situational eye-contact or in the words of a love conversation.

In his lectures on *Grundbegriffe der Metaphysik* (1929/30)[14] Heidegger wrote: ". . . first, moods are not beings, not things that somehow simply appear in the soul; second, moods are not that which is most inconstant and fleeting either, contrary to what people think."[15] If we can say rightly that *Stimmung*, mood, is not being, not anything, then we must accept the thought that being, something, or the thing is not the thing, but "the half-thing" (the technical term of Schmitz for night, sound, look etc.). ". . . [M]ood is not at all inside, in some sort of soul of the Other, and . . . it is not at all somewhere alongside in our soul. Instead we have to say, and do say, that mood imposes itself on everything. It is *not* at all '*inside*' in some interiority, only to appear in the flash of an eye. . . . Mood is not some being that appears in the soul as an experience, but the way of our being there with one another . . . mood is infectious."[16] Mood is atmosphere. ". . . moods are a fundamental manner, the *fundamental way in which Dasein is as Dasein.* . . . Moods are the fundamental ways in which we *find* ourselves *disposed* in such and such a way."[17]

An atmosphere is a total and holistic situation. If you say this restaurant has an Oriental atmosphere and a good mood, how does everything exist in this total atmosphere? Is it enough to accept the traditional thing-concept since Democritus and Aristotle to explain mood?

To elucidate the mode of existence of a thing, I must take a brief glance at the concept of thing in the philosophical tradition since Aristotle. The thing is conceived in this tradition as a normative subject of a sentence and this subject bears its contents and determinations as predicates. Husserl is not entirely free from the substanialistic viewpoint within the Aristotelian tradition. One can find this strand of thought in his posthumous work *Experience and Judgment*[18] that was edited by his renowned assistant Ludwig Landgrebe. Husserl explicated inner and outer horizons with concepts like substrate and determinations [*Substrat und Bestimmungen*]. The substrate is an individual thing that is thematized through the act of consciousness. Properly speaking, these substantialistic concepts go against the horizon of Husserl's thought since the substrate of determinations is not an ontological substance or *hypokeimenon* as being, but belongs to the dimension of appearance. According to the epoche of Husserl and the archaic skeptical school, to be is understood as to appear.[19]

Today however the natural scientists normally presuppose the Democritian tradition, accepting the atomistic way of thinking. For example, a leaf of a tree is in itself neither green nor yellow, but a substance without colors. The substance of a leaf has no color, no taste, no odor and no touch, and consequently no sensitive qualities. This leaf-substance with unlimited nothingness of determinations is the thing that the natural scientist presupposes. This scientific, atomistic, and molecular thing has no sensitive-affective qualities; it is posited as a philosophical substantial reality and as subject.

The substance is enough by itself (*a se*). One posits this *ens a se* as subject. But a leaf appears as something green in the phenomenal world. The leaf is green. In order to describe this appearance of the green leaf one must set up substance without its determination as a subject. One will therefore say that the leaf is subject and has its determination, "green," as its predicate. The determination belongs to the subject. This inherence of the determination in the subject = substance is the outline of modern philosophical thought. Thus one finds dualities such as substance/attribute or subject/predicate. But my question now can be stated as follows: Is the thing in "atmosphere" properly conceived in this mode of thinking?

Clearly, the thing in an atmosphere is not *a se*, for any thing appears as an essential component of the total atmosphere. A thing looks like a necessary element that lets atmosphere manifest itself. These elements cannot be lacking in order for an atmosphere to manifest itself. The thing is no longer the half-thing, or better, the radiation of things. For example, Chinese lanterns do not simply amount to tools to lighten a dark room. They must lose themselves in the original mood of a Chinese restaurant. You can do a simple experiment: Instead of Chinese lanterns, use genuine Central-African lanterns. That ruins the entire atmosphere. The Chinese lanterns manifest the genuine Chinese atmosphere in that restaurant.

What has happened in this case? This is my thesis: This phenomenon of the Chinese restaurant depends on the destruction of substantiality of things. Everything extends and reaches out of itself into the environmental atmosphere. This reaching-out-of-itself to the environmental world can be expressed by the words "radiation of things." Every light-tool has ecstasies in the total original Chinese atmosphere. These lanterns are unified with the holistic mood of the restaurant. Everything in the atmosphere shows itself in modes of appearance such that the content-determinations of the thing break its rigid substantiality. The thing has its radiation from itself into the atmosphere. A garden under the gloomy sky of the rainy season mirrors its indigo color like hydrangea flowers. The color breaks the substantiality of a flower and lets it shine from its inner horizon. The substantiality, and the inherence of the determinations of the thing in the substance/subject, are broken up into the mood surrounding the thing.

According to my interpretation of the concept of atmosphere in Gernot Böhme,[20] what he names the *Ekstase der Dinge* is the mode of thing-appearance in the tradition of Husserl-interpretation. Husserl's late thought on passivity converged on the stage of the thing-appearance. The thing does not appear as an individual substance, but only with and in horizons. The outer and inner horizons of a thing are the modes of its thing-appearance. Atmosphere is a kind of horizon. Therefore we can say that a thing appears in its mode of appearance; the thing comes to light in and through its modes of appearance. And this thing appears to me. The modes of appearance cannot avoid the relation to the seeing subject.[21]

Atmosphere is nothing but mood, air, and wind. Or what we call atmosphere is the modes of the appearance of world and thing, namely the horizons in which human beings who see and touch are intertwined with objects that are seen and touched. Atmosphere is the medium between the so-called subject and object. Through this medium[22] we can perceive the appearance of world and thing. Here we see the deep connection between atmosphere and the phenomenology of wind.[23]

Synaesthesia

Atmosphere appears to be dependent upon every sense. If you say atmosphere appears in relation to sight, then atmosphere appears only in a visible way. But it is obvious that atmosphere is not of the visible or of the tactile. Nor is it audible. Atmosphere is a *holon*, a *totum*. It is therefore not captured by any single sense. Atmosphere is the symphony of all sensations and is to be grasped through the orchestration of all senses. The mood of a café may be wonderful, but you cannot designate what part of the mood is good or better than anything else. This orchestration of all the participating senses is called synaesthesia in the broadest sense of the word. Atmosphere or mood is something corresponding to synaesthesia.

To understand atmosphere one must not analyze the orchestra of all the senses into its parts. The appearance of color in sight concords with or corresponds to other senses like touch or hearing.

Synaesthesia used to be analyzed as a problem of metaphor, as Roman Jakobson, the famous structural linguist, had done. For example, expressions like high sounds or low sounds are a kind of metaphor. Whereas sound is originally an acoustic appearance that has no spatial orientation, high and low depend upon a visual perspective in space, at the center of which is my viewpoint.

It should be noted that the linguistic solution to this problem is not adequate. We must consider what makes it possible for us to understand synaesthetic expressions in language as rhetorical expressions. Expressions like the Japanese "yellow voice" or the English, "soft voice," or the German "velvety music" do not have a visual yellow content, a tactile sensation of softness, or a tactile perception of velvet. But we can understand these metaphoric words. The understanding of these metaphorical expressions presupposes that there is a dimension of something common in sight and touch and sound.

What is this common 'something'? And how is it possible to bring this dimension to light? A sense in one dimension of sense will radiate to all other sense-realms. The tactile sense "velvety" mirrors the acoustic dimension. The color "black" carries us down to the Underworld. The color "black" is quiet, calm, and warm. Bringing one down, the black color does not jump up and down lightly or superficially. The heavy and calm touching tone which the color "black" shows is appropriate to a funeral. In his 9th Symphony, Beethoven denied the light-hearted and cheerful musical theme by using the heavy and dark sound from the contrabass and violoncello. The music of "Nein" must come out of the right wings of the orchestra.

The feeling of body states

This is where my last question arises: Where is the stage set for the mirroring and radiating interplay of the senses in different sense-regions? My answer runs as follows: the stage is one's body. But the body must be understood in the sense of "Leib" or lived-body, and not as an anatomical or physiological construction.

The original place of this synaesthesia is one's body and one's body-state.[24] When I see the color black, I have the sensation of stepping into the Underworld in my body-state. Being tired or overcome by Morpheus,

I have sometimes had the experience of my body being drawn under. When I see the color black, I represent to myself oppressive experiences. Do not mistake this as following the same line of explanation as the associationalist theory of psychology. Psychological associationalism supposes mental states in a consciousness that are divided into atomistic parts. With its artificial connection of atomistic mental states, this theory seems to explain the fact of oppressive experience before the black color. But I presuppose neither the mental states nor its atomistic conception.

This synaesthesia lies deep in body-states in which the body finds itself in its environment. The origin of synaesthesia is rooted in the modes of body-states. The holistic totality of atmosphere that expands to the end of world is grasped by the body-states of feeling-mood in which sensations emanate and mirror each other.

My body occupies the absolute place "here." Since from the viewpoint of qi, the body of everybody is identical with the wide and deep world, this "here" is not a point like a geometric point but has endless depth and width. The body swells to the end of the world. Body-states appear in the awareness of powers in the body: the awareness of hunger, freshness, fatigue or languor. I think in this sense there is a difference between the concept of the body of qi/ki and that of Husserlian thought.

The phenomenological concept of the body in the Husserlian sense lies in the consciousness of the spontaneity *"Ich-bewege-mich"* ("I move myself"). But the phenomenology of qi elucidates the base of this spontaneous concept of the body, the relation of the small body, my body and the large body, namely, the world. This relation is nothing but the appearance of qi and atmosphere that is the most passive and deep dimension of the body and world. You will find the same idea expressed in texts from Plato's *Timaeus* (78E–79A). Expiration and inspiration, the movement of wind (*pneuma*) between the world and me appears as the deepest dimension of Being.

Atmosphere shows itself as the pre-predicative and even evident *logos* of the world. It is Hermann Ammann who analyzed the relation of *logos* and the appearance of mood descriptively. Children will sing loudly in the dark because hearing his or her own voice allows the atmosphere or mood of solitude disappear. By singing a song or whistling, which is nothing but the activity of breath and qi, one escapes one's own solitude; song is a language. In this point of view there is no monologue in the true sense but, as Ammann says: "If I do not speak with a human interlocutor, I can still speak to everything that surrounds me, yes, even partly to my own self ."[25] The unity of atmosphere with language or pre-logical *logos* of world is elucidated by him as chorus language: Many people are waiting for a train on the platform in the stormy wind and rain. The delayed train, however, does not come, and the irritating mood is infectious. One person making noise will be accepted by the others implicitly, and another will explain his complaints explicitly. This atmosphere on the platform and its language are experienced intersubjectively as inner language form that Ammann names "chorus language".

Atmosphere appears in the modes of filling the world. Atmosphere, on the one hand, reveals the whole world as proto-logical truth. On the other hand, atmosphere attacks and grasps my body.

The binding point of both are the states of my body. My body transcends the contoured thing-body and extends out to the world. My body too has radiations in the atmosphere that is the world-appearance. Atmosphere as mood grasps me, and I am aware of this atmosphere in my body-states. Toko Fujita of the Mito-School explained this fact: "Qi/ki gathers and concentrates in me from the all sides of the world." The phenomenology of qi/ki and wind elucidates the fact that qi/ki fills the body and flows over the world at the same time.

Notes

1. Dogen, *Shobo-Genzo*, the third book Genjyo-Koan, Iwanami Universal-Bibliothek, Volume One, 87.
2. Yasushi Aizawa, *Shinron und Itekihen*, Iwanami-Universal-Bibliothek, 260, 272.
3. Aizawa, *Shinron und Itekihen*, 272.
4. To the concept of intercultural invariance see my article "Translation as a cultural-philosophical problem: Towards a phenomenology of culture, in *The Monist*, Vol. 78, No. 1, 1995.
5. Plato, *Timaeus*, 78E–79A.
6. C. Baudelaire, Le Gouffre, *Les Fleurs du Mal*, Editions Garnier, Paris 1961, p. 189, especially the notebook of Baudelaire, p. 448. English translation by Lewis Piaget Shanks, *Les Fleurs du Mal: Baudelaire Complete Poems*: " – Alas! I see abysses everywhere:/Dreams, action words! and o'er my lifted hair/Full oft I feel the wind of Horror glide."
7. Edmund Husserl, the founder of the phenomenology, was able to open the fundamental dimension; this is nothing other than the fact that human beings are deeply and definitively rooted in the world. Concerning his phenomenology of passivity it suffices to say the following: In the phenomenology of active constitution, I have the object constituted before me. Generally speaking, objectivity presupposes that the developed ego constitutes formal unification from material complexity as the object as such. Therefore, the crux of the problem in passive phenomenology concerns the realm in which the ego is not yet developed enough as ego, namely, the sphere prior to all activity of the ego.

In the passive sphere, the pregivenness prior to the apperception of a unified object as such appears as *hyle* in the domain of the passivity. This pregivenness, *hyle*, that works upon the ego Husserl calls "*Reiz*," allure, excitation. With this concept, Husserl has in mind the affection of the pre-egoic world on the developing ego. Like the landscape seen by the extremely shortsighted eyes, the unity-formation is withdrawn. Only the *hyle*, namely, the extension of colors or sounds appear in the phenomenal field. The phenomenology of passivity in Husserl allows us to see this dimension as something structured

334

with logical meaning in which the pre-objective and the pre-subjective interpenetrate. In the background of all predicative logical truth there must be a pre-predicative world which is evidently opened up in the pre-linguistic level. It is the body that plays such an eminent in opening up this prelinguistic world. "The world is constituted in such a way that in it, even the ego emerges as embodied." ["Die Welt ist so konstituiert, daß in ihr selbst das Ich als verleiblichtes auftritt."]. Edmund Husserl, *Phänomenologie der Intersubjektivität*, Husserliana Vol. XV (The Hague: Nijhoff, 1973), 287.

Concerning Husserl's notion of passivity, see Edmund Husserl, *Analysen zur passiven Synthesis*, Hua. XI. S 101–102. [English translation by Anthony J. Steinbock, *Analyses Concerning Passive and Active Synthesis: Lectures on Transcendental Logic, Husserliana Collected Works* (Boston: Kluwer, forthcoming)]. See my article: *Das Hervorgehen der Sprache (Logos) aus der natürlichen Welterfahrung*, in *Japanische Beiträge zur Phänomenologie*, Alber: Freiburg i.Br. -München 1984. I have thematized there the relation of atmosphere in Schmitz and passivity in Husserl.

8. Martin Heidegger, *Sein und Zeit* (Tübingen: Niemeyer, 1927), 136: "Die Stimmung kommt weder von "Außen" noch von "Innen", sondern steigt als Weise des In-der-Welt-seins aus diesem selbst auf. Die Stimmung überfällt." English translation by John Macquarrie and Edward Robinson, *Being and Time* (New York: Harper and Row, 1962), 176: "A mood assails us. It comes neither from 'outside' nor from 'inside', but arises out of Being-in-the-world, as a way of such Being."

9. Hermann Schmitz, *System der Philosophie*, Bd. III./2., (Bonn: Bouvier, 1969)

10. Klaus Held, "Die gegenwärtige Lage der Philosophie – Heideggers Phänomenologie der Grundstimmungen, in *Zur philosophischen Aktualität Heideggers*, eds., Dietrich Papenfuss und Otto Pöggeler, Ffm, 1990. English translation by Anthony J. Steinbock, "Fundamental Moods and Heidegger's Critique of Contemporary Culture," in *Reading Heidegger: Commemorations*, ed., John Sallis (Bloomington: Indiana University Press, 1993), 286–303.

11. Heidegger, *Sein und Zeit*, 142: "Verstehen ist immer gestimmtes."

12. Heidegger, *Being and Time*, 177 (translation slightly modified). Heidegger, *Sein und Zeit*, 137–138: "In der Befindlichkeit liegt existenzial eine erschließende Angewiesenheit auf Welt, aus der her Angehendes begegnen kann."

13. Emile Durkheim, *Les formes élémentaires de la vie religieuse* (Paris: PUF, 1968); following the Japanese Translation, Iwanami-Universal-Bibliothek, Vol. I. p. 378–380.

14. Heidegger, M.: Gesamtausgabe Bd. 29/30. *Die Grundbegriffe der Metaphysik* S 98 ff.

15. Martin Heidegger, *The Fundamental Concepts of Metaphysics: World, Finitude, Solitude*, trans., William McNeill and Nicholas Walker (Bloomington: Indiana, 1995), 65–66 (translation slightly modified). "Die Stimmungen sind erstens kein Seiendes, kein Etwas, das irgendwie in der Seele nur vorkommt; Stimmungen sind zweitens ebensowenig das Unbeständigste und Flüchtigste, wie man meint."

16. Heidegger, *Fundamental Concepts*, 66–67 (translation slightly modified). Heidegger, GA 29/30, 100–101: "Die Stimmung ist sowenig darinnen in irgendeiner Seele des anderen und sowenig auch daneben in der unsrigen, daß wir viel eher sagen müssen und sagen: Diese Stimmung legt sich nun über alles, sie ist gar nicht darinnen in einer Innerlichkeit und erscheint dann nur im Blick des Auges; . . . Die Stimmung ist nicht ein Seiendes, das in der Seele als Erlebnis vorkommt, sondern das Wie unseres Miteinander-Daseins" "Die Stimmung steckt an."

17. Heidegger, *Fundamental Concepts*, 66–67 (translation slightly modified). Heidegger, GA 29/30, 100–101: "Die Stimmung ist . . . die Grundweise, wie das Dasein als Dasein ist. . . . Die Stimmungen sind die Grundweisen, in denen wir uns so und so befinden."

18. Edmund Husserl, *Erfahrung und Urteil*, ed., Ludwig Landgrebe (Meiner: Hamburg 1972), §§26–29. English translation by James S. Churchill and Karl Ameriks, *Experience and Judgment*, ed., Ludwig Landgrebe (Evanston: Northwestern University Press,, 1973).

19. See my book, *The Logos of Phenomenon*, (Tokyo, 1986), 14–47.

20. See Gernot Böhme, *Atmosphäre*, 1995. Since there are some naturalisitic moments in his concept of atmosphere, I will purify his concept of them and take him to task in the following description.

21. Cf. my book: *The Logos of Phenomenon*, Tokyo 1986.

22. Aristotle said in his book "On the soul": *"aisthanometha ge panton dia tou mesou."* We perceive all things through a medium. *De Anima*, 423 b 6–7. Loeb Classical Library, 1975, 132–133.

23. According to a Buddhistic school in India, *vaisesika*, time is conceived as analogous to wind, an element among four fundamental elements (earth, fire, water, and wind). Since they are difficult to objectify and because they are movements that one cannot see, wind and time are not to be seen, but felt. E. Ejima, "Time theory of the Mahayana-Buddhism," in *Bukkyo-Shiso*, vol. 1. (Tokyo: Riso-sha 1974, 1980), 236 f.

24. I understand under the concept of body-state that of the *"Zustand" des Leibs. Zustand* is non-intentional way of appearance of a body. Cf. Klaus Held: *Heraklit Parmenides und der Anfang von Philosophie und Wissenschaft*, Berlin-New York 1980. P. 371.

25. "Spreche ich nicht zu einem menschlichen Hörer, so kann ich doch zu allem sprechen, was mich umgibt, ja selbst zum Teilen meines Ich." Hermann Ammann, *Die Menschliche Rede* (Darmstadt: 1974, fourth edition), 168.

Note about the author

Tadashi Ogawa (born 1945 in Osaka, Japan) completed doctoral studies at Kyoto University (1974), was a research fellow at the Husserl-Archives in the University of Cologne, Germany (1975–1976), and became professor of European Philosophy at Hiroshima University (1978 where he was promoted to full professor in 1990). He was a Humboldt-Stipendiat at Bergische Universität Wuppertal and Christian-Albrechts University Kiel (1982–1983), and took a position as Professor of Philosophy at Kyoto University (1991). In addition to articles that have their basis in the phenomenological tradition, his books include: *The Logos of Phenomenon: A System of Structuralistic Phenomenology* (Tokyo, 1986); *Phenomenology and Cultural Anthropology* (Tokyo, 1989); *Phenomenology and Structuralism* (Tokyo, 1990); and *Structure to Liberty: A Study of the History of Political Philosophy*.